Virus Attachment and Entry into Cells

Virus Attachment and Entry into Cells

Proceedings of an ASM Conference Held in Philadelphia, Pennsylvania,
10–13 April 1985

Editors:

Richard L. Crowell and
Karl Lonberg-Holm

American Society for Microbiology
Washington, D.C. 1986

Library of Congress Cataloging-in-Publication Data

Main entry under title:

Virus attachment and entry into cells.

  Conference sponsored by the American Society for Microbiology.
  Includes bibliographies and index.
  1. Viruses—Receptors—Congresses. 2. Cell receptors—Congresses. 3. Adsorption (Biol-
ogy)—Congresses. 4. Host-virus relationships—Congresses. I. Crowell, Richard L. II.
Lonberg-Holm, Karl. III. American Society for Microbiology. [DNLM: 1. Cell Mem-
brane—congresses. 2. Membrane Proteins—congresses. 3. Receptors, Virus—congresses.
QW 160 V8198 1985]
QR469.V57 1986   576v'.64   85-28731

ISBN 0-914826-90-5

Contents

Preface vii

Introduction and Overview. KARL LONBERG-HOLM AND RICHARD L. CROWELL .   1

Virus Attachment Proteins

Nature and Function of the Reovirus Hemagglutinin in Cell and Tissue Tropism. RHONDA BASSEL-DUBY, ANULA JAYASURIYA, AND BERNARD N. FIELDS 13

Location of Four Neutralization Antigens on the Three-Dimensional Surface of a Common-Cold Picornavirus, Human Rhinovirus 14. ROLAND RUECKERT, BARBARA SHERRY, ANNE MOSSER, RICHARD COLONNO, AND MICHAEL ROSSMANN... 21

Monoclonal Antibodies as Probes of the Poliovirus Surface and of the Cellular Receptor. P. D. MINOR, D. EVANS, M. FERGUSON, J. W. ALMOND, AND G. C. SCHILD....................................... 28

Anti-Idiotopic Antibodies against Anti-Coxsackievirus Type B4 Monoclonal Antibodies. PATRICK R. MCCLINTOCK, BELLUR S. PRABHAKAR, AND ABNER LOUIS NOTKINS 36

Biochemical Characterization of Polyomavirus-Receptor Interactions. R. A. CONSIGLI, G. R. GRIFFITH, S. J. MARRIOTT, AND J. W. LUDLOW....... 44

Viral Membrane Fusion Proteins. JUDY WHITE, ROBERT DOMS, MARY-JANE GETHING, MARGARET KIELIAN, AND ARI HELENIUS 54

Monoclonal Antibodies as Probes for Influenza Virus Hemagglutinin Structure and Function during the Infectious Cycle. JONATHAN W. YEWDELL, ALEX TAYLOR, ANDREW CATON, WALTER GERHARD, AND THOMAS BACHI... 60

Membrane-Active Peptides of the Vesicular Stomatitis Virus Glycoprotein. RICHARD SCHLEGEL 66

Identification and Analysis of Biologically Active Sites of Herpes Simplex Virus Glycoprotein D. ROSELYN J. EISENBERG AND GARY H. COHEN 74

Protein-Mediated Fusion of Viral and Cellular Membranes. FRANK R. LANDSBERGER AND PRAVINKUMAR B. SEHGAL 85

Role of the Viral Envelope in Leukemia Caused by Friend Murine Leukemia Virus. ALLEN OLIFF 91

Cellular Receptors

Purification of a Membrane Receptor Protein for the Group B Coxsackieviruses. JOHN E. MAPOLES, DAVID L. KRAH, AND RICHARD L. CROWELL 103

Characterization of the Cellular Receptor Specific for Attachment of Most Human Rhinovirus Serotypes. RICHARD J. COLONNO, JOANNE E. TOMASSINI, PIA L. CALLAHAN, AND WILLIAM J. LONG.......................... 109

Encephalomyocarditis Virus Attachment. GRAHAM P. ALLAWAY, INGRID U. PARDOE, AMIR TAVAKKOL, AND ALFRED T. H. BURNESS 116

Nature of the Interaction between Foot-and-Mouth Disease Virus and Cultured Cells. BARRY BAXT AND DONALD O. MORGAN 126
Structural Characterization of the Mammalian Reovirus Receptor. MAN SUNG CO, GLEN N. GAULTON, JING LIU, BERNARD N. FIELDS, AND MARK I. GREENE 138
Biological Implications of Influenza Virus Receptor Specificity. JAMES C. PAULSON, GARY N. ROGERS, JUN-ICHIRO MURAYAMA, GLORIA SZE, AND ELAINE MARTIN 144
Nature of the Rabies Virus Cellular Receptor. WILLIAM H. WUNNER AND KEVIN J. REAGAN 152
Biological Significance of the Epstein-Barr Virus Receptor on B Lymphocytes. GLEN R. NEMEROW, MARTIN F. E. SIAW, AND NEIL R. COOPER 160

Penetration and Uncoating

Entry Mechanisms of Picornaviruses. SJUR OLSNES, INGER HELENE MADSHUS, AND KIRSTEN SANDVIG 171
Uncoating of Vesicular Stomatitis Virus. JOHN LENARD 182
Pathway of Adenovirus Entry into Cells. PREM SETH, DAVID FITZGERALD, MARK WILLINGHAM, AND IRA PASTAN 191
Requirements for Initial Adenovirus Uncoating and Internalization. EINAR EVERITT, ULLA SVENSSON, CLAES WOHLFART, AND ROBERT PERSSON 196
Prospects for Antiviral Agents Which Modify the Pathway of Infection by Enveloped Viruses. ARI HELENIUS, MARGARET KIELIAN, JUDY WHITE, AND JÜRGEN KARTENBECK 205

Author Index 211
Subject Index 212

Preface

A conference on "Virus Attachment and Entry Into Cells," sponsored by the American Society for Microbiology, was held 10–13 April 1985 in Philadelphia. The goal of this conference was to provide an informal examination of current work on the early events of virus infection of animal and human cells. As far as we are aware, this was the first international conference on this subject. The growing interest in this area was attested by the attendance of about 270 researchers, from 15 countries and 29 states of the U.S.A. With only 2½ days for presentations, it was not possible to cover the entire field exhaustively, and omissions of some subjects and individual presenters became apparent after the planning was completed. Despite this inevitable limitation, the collected presentations are representative of current virus-receptor research, a field which forms a bridge between the disciplines of microbiology, cell biology, and molecular biology.

The 24 chapters are organized arbitrarily into three areas: Virion Attachment Proteins, Cellular Receptors, and Penetration and Uncoating of Viruses. In the first chapter of this volume, we attempt to provide a brief overview of the early events in virus infection. The minireview format of most of the contributions in this volume as well as our introductory chapter has precluded a comprehensive literature citation of many meritorious published works. More detailed references to the literature can be found in the reviews referred to in each chapter.

We are grateful to the following for generous support which helped to make this conference possible.

Abbott Laboratories
American Cyanamid Company
Analytab Products
Bristol-Myers Company
Burroughs Wellcome Company
DuPont Biomedical Products
E.I. Du Pont De Nemours & Company
Hoffmann La Roche, Inc.
Merck Sharp & Dohme Research Laboratories
Schering Corporation
Smith Kline & French Laboratories
Sterling-Winthrop Research Institute
Wyeth Laboratories
National Science Foundation (grant DCB-8414982)
National Institute of Allergy and Infectious Diseases (Public Health Service grant AI-22015-01)

We also are indebted to Rita Vogler of the ASM Meetings Department, who coordinated the administrative and logistic details for this conference in an extraordinary manner, and to numerous volunteers from the Hahnemann University Department of Microbiology and Immunology. For her capable help in editing and retyping manuscripts, we thank Margie Abrams. We gratefully acknowledge the cooperative spirit shown by the presenters in submitting their manuscripts and regret that the abstracts from the 41 posters presented at the conference could not also be published.

KARL LONBERG-HOLM
Department of Microbiology and Immunology
Hahnemann University School of Medicine
Philadelphia, Pennsylvania 19102

Central Research and Development Department
E. I. duPont de Nemours & Co.
Experimental Station
Wilmington, Delaware 19898

RICHARD L. CROWELL
Department of Microbiology and Immunology
Hahnemann University School of Medicine
Philadelphia, Pennsylvania 19102

Introduction and Overview

KARL LONBERG-HOLM[1,2] AND RICHARD L. CROWELL[1]

Department of Microbiology and Immunology, Hahnemann University School of Medicine, Philadelphia, Pennsylvania,[1] and Central Research and Development Department, Experimental Station, E. I. DuPont de Nemours & Company, Wilmington, Delaware 19898[2]

We are witnessing a period of rapid progress in understanding the details of the early steps in virus infection of cells. This is the result of several new biological insights and biochemical technologies, applications of which are described in the collected papers of this symposium. Discrete virion attachment proteins have been identified in many viruses, and their functional homologs are being identified in regions of the capsids of the small nonenveloped picornaviruses. In the latter case, there is surprising evidence from X-ray crystallographic analysis that the receptor binding site is a cleft which is shielded from antibodies that might otherwise react with group-specific receptor-recognizing determinants. Host-cell receptors are being isolated, often with the use of monoclonal antibodies, but also by application of other techniques. It can be anticipated that these receptors will soon be biochemically characterized and their cDNA will be sequenced. The molecular details of the function of virion fusion proteins, which interact with the cell membrane during penetration, are being elucidated. Many, but not all, viruses appear to use the endosomal pathway for adsorptive endocytosis, and in these cases, inhibitors of endosomal acidification also inhibit infection of host cells. The capsid or envelope proteins of these viruses appear to be able to undergo charge and conformational alterations at the acid pH of the endosome which lead to viral membrane penetration and entry into the cytoplasm.

VIRION ATTACHMENT PROTEINS

The initial event in the life cycle of a virus is attachment to specific receptors on a host cell (27, 46). This event is often a major determinant of virus tropism in pathogenesis (8, 43). There also may be nonspecific binding, but this is nonsaturable and generally does not lead to productive infection. However, there may be exceptions for some viruses, and it is possible that nonspecific attachment may lead to infection (Lenard, this volume). In the case of the flaviviruses, attachment specificity may be provided by host antibodies attached to virions, since recognition of Fc receptors by the Fc portion of antibodies can lead to infection of leukocytes (13).

It is likely in many cases that there is more than one type of virion determinant and each determinant recognizes a different host-cell receptor (40, 47). Of course these determinants may be located on the same molecule on the capsid or envelope of the virion. For example, hemagglutinating variants of

1

group B coxsackieviruses appear to acquire a new receptor specificity for infection of RD cells while retaining the ability to compete with parental virus for HeLa-cell receptors (40). Similarly, a poliovirus host-range variant selected for growth in murine cells (which lack receptors for the prototype virus) appears to have acquired a second specificity, since it also grows in human cells (38). Structures which recognize cellular receptors for reoviruses, picornaviruses, and papovaviruses are presented first in this volume. Virion attachment proteins (VAP) are specialized structures in many viruses; for example, the sigma 1 protein of the reoviruses projects from the vertices of the virions (Bassel-Duby et al., this volume). A single amino acid change in the sigma 1 polypeptide (selected under pressure from monoclonal neutralizing antibodies) was associated with changes in tissue tropism and pathogenicity in the host animal, presumably as a result of a concomitant altered receptor specificity (Bassel-Duby et al., this volume). Single amino acid substitutions in the influenza virus hemagglutinin are also associated with altered receptor specificity (Paulson et al., this volume).

The VAP of enveloped viruses are often characterized as viral hemagglutinins. Morphologically they are seen as flexible "spikes" on the virions. The elongated shape of these glycoproteins may aid attachment by reducing the decrease in freedom of motion or entropy following attachment (23). However, these viral membrane proteins have functions in addition to recognition of either erythrocyte or host-cell receptors. For example, the role of the influenza virus hemagglutinin protein in fusion and penetration of the host-cell membrane is now beginning to be characterized in molecular detail (White et al., this volume; Yewdell et al., this volume), and it is possible to make detailed comparisons with fusion proteins of paramyxoviruses where the VAP is a separate molecule. Fragments of the vesicular stomatitis VAP were found to have biological activity as hemolysins (Schlegel, this volume). The mechanisms by which hemagglutinins cause either hemolysis or fusion of viral and host-cell membranes remains controversial. An attempt to relate their biochemical activity to that of lipid transport proteins has been made (Landsberger and Sehgal, this volume), but other mechanisms also must be considered.

In contrast, the picornavirus receptor recognition determinant may comprise neighboring regions within one of the capsid polypeptides, VP1 (Rueckert et al., this volume). The evidence for this comes from detailed comparisons of the X-ray crystallographic structure of human rhinovirus 14 and sequence data on antibody-resistant mutants. In this model, a narrow cleft in the capsid wall recognizes and binds the cellular receptor. However, antibodies cannot fit into this determinant-containing cleft, although they may be able to recognize the edge of the cleft and block entry of the cellular receptor. Thus, members of a "receptor family" of viruses need not share a common antigen on the virion surface. Furthermore, anti-idiotype antibodies to picornavirus-neutralizing antibodies will not recognize cellular receptors. Preliminary attempts to use anti-idiotype antibodies to react with receptors for coxsackievirus type B4 (McClintock et al., this volume) and for foot-and-mouth disease virus (Baxt and Morgan, this volume) have failed. This suggests that the application of the "internal image" theory of antibody specificity to viral receptors as for reovirus (Co et al., this volume) may not be possible with picornaviruses.

The specificity of the VAP appears to play an important role in the pattern of virus pathogenesis (8; Bassel-Duby et al., this volume). VAP specificity has been explored by biochemical methods for rabies virus (Wunner and Reagan, this volume) and by genetic methods for a murine leukemia virus (Oliff, this volume). The VAP of enveloped viruses and reoviruses also present a major target for neutralizing antibodies. RNA genomic viruses like picornaviruses (Minor et al., this volume; McClintock et al., this volume), influenza virus (Yewdell et al., this volume), reoviruses (Bassel-Duby et al., this volume), and acquired immune deficiency syndrome (AIDS) virus (39) may all be able to circumvent neutralizing antibodies by mutation of the major antigenic sites of their external proteins (17). A similar strategy may be used by DNA viruses, although at a lower frequency of mutation. Nevertheless, there is reason to hope that conserved antigenic domains of the herpesvirus VAP (glycoprotein D) may be used for design of effective virus type-common vaccines, as discussed by Eisenberg and Cohen (this volume).

CELLULAR RECEPTORS

The virus usually, but not always, uses a host-cell intrinsic membrane glycoprotein for its receptor. Table 1 summarizes the tentative identities of receptors for 15 different viruses. Other information also may be available. For example, the chromosomal location of the poliovirus receptor is known (33); it is antigenically and functionally present only on human and primate cells (Minor et al., this volume). Receptors usually differ from virus to virus, although unrelated viruses may "seize upon" the same receptor; for example, adenovirus type 2 shares the receptor for coxsackieviruses of group B (24). In general, however, viruses within a family evolve while conserving their receptor specificity. It would be disadvantageous for the virus to retain this specificity if it required retention of a common neutralizing antigen. One way around this problem may be the strategy already mentioned for picornaviruses, where the recognition site for the receptor is sterically shielded from antibodies; at least 75 human rhinovirus serotypes share the same receptor and presumably have conserved receptor-recognizing determinants (Colonno et al., this volume). Another strategy might be adaptation to "nonspecific" attachment, as suggested for rhabdoviruses (Lenard, this volume).

It may be speculated that virus receptors not only act as points of attachment for the virus but also contribute to subsequent steps in the infection pathway (8). For example, the receptors may be strategically located for facilitating virus entry into the cell (37). Since viruses are repeating subunits, and the plasma membrane is fluid at physiological temperatures, the virus-receptor complex is composed of many cellular receptor units combined in one multivalent cellular receptor site (23). One of the consequences of this is that the individual bonds between VAP and receptor subunits may be weak, while the avidity of the multivalent complex may be high. A second consequence is that because of cooperative interactions (Everitt et al., this volume) the properties of the virus complexed in the cellular receptor site may be very different from those of the isolated components. For example, isolated cellular receptors may bind very poorly to virions or may not be able to participate in uncoating in the way that

TABLE 1. Tentative assignment of cellular structures as receptors for viruses

Virus	Organelle or molecule	Approx mol wt	Cell type	Reference
Enveloped				
Semliki Forest	H-2K and H-2D	44,000	Multiple	15, 36
Sindbis	Catecholaminergic neurotransmitter (?)	(?)	Skeletal muscle	47
Epstein-Barr	Complement receptor 2	145,000	B lymphocytes	12, 34
	Class II antigen	60,000	B lymphocytes	41
LAV/HTLV-III	CD4 (T4) antigen	55,000	T lymphocytes	11, 19
Murine leukemia (Rauscher)	(?)	10,000	Fibroblasts	22
Rabies	Acetylcholine receptor (?)	292,000	Skeletal muscle	5, 52
Vesicular stomatitis	Phospho- or glycolipid	<10,000	Fibroblasts	2, 42
Lactate dehydrogenase	Ia antigen	30,000	Macrophages	18, 21
Influenza A	Glycophorin A, sialyloligosaccharides	84,000	Erythrocytes	16, 30
Sendai	Gangliosides	<10,000	Multiple	31
Nonenveloped				
Reovirus T3	Beta-adrenergic hormone receptor	67,000	Neurons, lymphs, L-cells	6
Encephalomyocarditis	Glycophorin A	84,000	Erythrocytes	4
Group B coxsackievirus	Rp-a (?)	49,500	HeLa	29
Rhinovirus	(?)	80,000	HeLa	—[a]
Adenovirus	(?)	42,000	HeLa	45

[a] Tomassini and Colonno, submitted for publication.

is possible in the intact complex which contains membrane lipids as well as receptors. However, for some viruses the capsid may be labelized within the isolated receptor complex, as suggested for uncoating of coxsackieviruses (9).

Since there are usually only about 10^4 to 10^5 receptor units per host cell (25), it has been almost prohibitively difficult to obtain sufficient material for biochemical characterization (Consigli et al., this volume). One strategy around this has been to study erythrocyte hemagglutination receptors. Since glycophorin is the major external erythrocyte membrane protein, it is not surprising that it is the receptor for many hemagglutinating viruses. Burness and co-workers have shown that a specific site on glycophorin A is the erythrocyte receptor for encephalomyocarditis virus (Allaway et al., this volume). This site contains a specific amino acid sequence in addition to the carboxylic acid group of only one of the sialic acid residues and is located relatively close to the anchoring point of this sialoglycoprotein on the cell membrane (Allaway et al., this volume).

Different strategies have been used for isolation of picornavirus host-cell receptors. Mapoles et al. (this volume) achieved greater than 10^4-fold purification of the group B coxsackievirus receptor by isolation of the coxsackievirus-receptor complex from infected cell membranes. They were able in this way to obtain sufficient material for preliminary characterization and demonstration of biological activity. However, this approach may not be of general application since many virus-receptor complexes are unstable during purification.

A more generally applicable strategy uses antireceptor antibodies for receptor purification. It is possible to screen for monoclonal antibodies to receptors by their ability to protect cells against virus attachment and infection (Minor et al., this volume; Colonno et al., this volume). However, monoclonal antireceptor antibodies are usually generated at a very low frequency. For example, only one in several thousand clones was found to be specific for the major group of human rhinoviruses (Colonno et al., this volume) or for coxsackieviruses of group B (7). Minor and co-workers (this volume) have attempted isolation and characterization of the receptor for poliovirus by using monoclonal antibodies, and there has been considerable progress in isolation of the 80-kilodalton major human rhinovirus receptor (J. E. Tomassini and R. J. Colonno, submitted for publication). It will be interesting to compare the structure of the cellular receptor with the detailed structure of the receptor-binding cleft derived from X-ray crystallographic data on human rhinovirus type 14 (Rueckert et al., this volume).

A rather clever approach to obtaining antireceptor antibody involves use of anti-idiotype antibodies to the paratope of an anti-VAP antibody. Although this approach may not be generally applicable, as already discussed above, it has been used for isolation of the reovirus type 3 receptor by Co et al. (this volume). Polyclonal anti-idiotype antibodies to a mouse monoclonal anti-virion sigma 1 antigen effectively bound to the cellular receptor. Interestingly, this isolated 67-kilodalton receptor, following sodium dodecyl sulfate-polyacrylamide gel electrophoresis and immunoblotting, bound not only antibody but also virus.

In some cases, the monoclonal anti-virus receptor antibody has been produced for other purposes. For example, the OKT4 monoclonal antibody appears to react with the receptor for HTLV-III/LAV (AIDS virus) (11,19), while OKB7 monoclonal antibody to the receptor for complement C3d also blocks attachment to lymphocytes of Epstein-Barr virus, as reviewed in detail by Nemerow et al. (this volume). We may expect rapid progress in the characterization and sequencing of many viral receptors in the near future as a result of the use of monoclonal antibodies. It also can be expected that sequence data will rapidly lead to cloning of the receptor DNA and the unfolding of many details of the synthesis, regulation, and function of these important cell-surface molecules.

PENETRATION AND UNCOATING OF VIRUSES

There has been a dramatic increase in our understanding of the mechanism of entry of viruses into host cells during the past 5 years. Much of this can be attributed to two conceptual developments: appreciation of the role of clathrin-coated pits in adsorptive endocytosis (1, 37), and the realization that endosomal acidification (32, 35, 37, 48) changes the properties of many internalized enveloped virions in such a way that they are able to react directly with the cell membrane's lipid bilayer (14, 44, 49–51; White et al., this volume). Enveloped viruses enter the cell in either clathrin-coated pits or other close-fitting vesicles (Helenius et al., this volume), and in most cases the viral envelope then fuses with the endosomal membrane and the nucleocapsid is injected directly into the cytoplasm.

Elucidation of this endosomal pathway has depended heavily upon the use of chemical inhibitors to prevent acidification of the endosomes. Helenius and co-workers and Pastan and his group have provided the driving force for much of this work. A discussion of future prospects for antiviral agents as inhibitors of the early steps of penetration and uncoating is presented herein by Helenius and co-workers.

There is less of a consensus over the mechanism of the entry and uncoating of nonenveloped viruses. The polemical debate over penetration by "viropexis" (10) or direct virus penetration (26) is not yet resolved. According to the viropexis model, virions are taken up in endosomal vesicles and then gain entry directly into the cytoplasm after virus-induced dissolution of the vesicle. Seth and co-workers (this volume) present an up-to-date version of this model for the entry of adenoviruses. Their data show that the adenovirus particle gains lipophilic properties at endosomal pH, suggesting that the modified virion fuses with or intercalates into the endosomal vesicular membrane, leading to its rupture. Brown and Burlingham also showed ultrastructural evidence for intercalation of adenovirus particles into host-cell membranes more than a decade ago (3).

The mechanism of penetration and uncoating of picornaviruses has been debated extensively over the past 10 years (8, 27). Olsnes and co-workers (this volume) have suggested that the poliovirus and rhinovirus pathway in early infection resembles that of many of the enveloped viruses. This model is summarized as follows. The virus attaches to specific receptors, and the complexes are taken into endosomes and acidified to pH below 6. The virion is altered by the acid pH to an amphipathic intermediate particle (which may be equivalent to or a precursor for the well-studied modified virions referred to as A particles), inserts itself into the vesicle membrane, and releases genomic RNA into the cytoplasm. Although the evidence for this model is incomplete, it is attractive because it explains a few puzzling observations. For example, it has been recognized for some years that poliovirus and rhinovirus type 2 A particles are amphipathic, and that this could be a clue to the mechanism for viral penetration and uncoating (25). Since acidification of the endosome requires ATP, the pH does not increase as usual within the endosomes of poisoned cells, and this explains earlier observations (28) that "eclipse" of cell-associated poliovirus and human rhinoviruses (loss of infectivity due to partial uncoating) is slowed by metabolic poisons. It was also found in the earlier work that metabolic poisons reduced the availability of rhinovirus receptors on the cell surface. This observation is viewed (Olsnes et al., this volume) as due to a block in the recycling of these receptors to the cell surface from the improperly acidified endosomes (see also reference 20).

In summary, the collected papers of this conference present a broad and useful introduction to current problems in the study of the early events of virus-cell interaction. They point the way to future discoveries and technological advances and suggest that the rate of progress in this field will increase substantially in the next few years.

LITERATURE CITED

1. **Anderson, R. G. W., M. S. Brown, and J. L. Goldstein.** 1977. Role of the coated endocytic vesicle in the uptake of receptor-bound low density lipoprotein in human fibroblasts. Cell **10**:351–364.
2. **Bailey, C. A., D. K. Miller, and J. Lenard.** 1984. Effects of DEAE-dextran on infection and hemolysis by VSV. Evidence that nonspecific electrostatic interactions mediate effective binding of VSV to cells. Virology **133**:111–118.
3. **Brown, D. T., and B. T. Burlingham.** 1973. Penetration of host cell membranes by adenovirus 2. J. Virol. **12**:386–396.
4. **Burness, A. T. H., and I. U. Pardoe.** 1983. A sialoglycopeptide from human erythrocytes with receptor-like properties for encephalomyocarditis and influenza viruses. J. Gen. Virol. **64**:1137–1148.
5. **Burrage, T. G., G. H. Tignor, and A. L. Smith.** 1985. Rabies virus binding at neuromuscular junctions. Virus Res. **2**:273–289.
6. **Co, M. S., G. N. Gaulton, B. N. Filedls, and M. I. Greene.** 1985. Isolation and biochemical characterization of the mammalian reovirus type 3 cell-surface receptor. Proc. Natl. Acad. Sci. USA **82**:1494–1498.
7. **Crowell, R. L., A. K. Field, W. A. Schlief, W. L. Long, R. J. Colonno, J. E. Mapoles, and E. A. Emini.** 1986. Monoclonal antibody that inhibits infection of HeLa and rhabdomyosarcoma cells by selected enteroviruses through receptor blockade. J. Virol. **57**:438–445.
8. **Crowell, R. L., and B. J. Landau.** 1983. Receptors in the initiation of picornavirus infections, p. 1–42. *In* H. Fraenkel-Conrat and R. R. Wagner (ed.), Comprehensive virology, vol. 18. Plenum Publishing Corp., New York.
9. **Crowell, R. L., and J. S. Siak.** 1978. Receptor for group B coxsackieviruses: characterization and extraction from HeLa cell plasma membranes. Perspect. Virol. **10**:39–53.
10. **Dales, S.** 1973. The early events in cell-animal virus interactions. Bacteriol. Rev. **37**:103–135.
11. **Dalgleish, A. G., P. C. L. Beverley, P. R. Clapham, D. H. Crawford, M. F. Greaves, and R. A. Weiss.** 1984. The CD4 (T4) antigen is an essential component of the receptor for the AIDS retrovirus. Nature (London) **312**:763–767.
12. **Fingeroth, J. D., J. J. Weiss, T. F. Teddler, J. L. Strominger, A. Biro, and D. T. Fearon.** 1984. Epstein-Barr virus receptor of human B lymphocytes is the C3d receptor CR2. Proc. Natl. Acad. Sci. USA **81**:4510–4514.
13. **Gollins, S. W., and J. S. Porterfield.** 1984. Flavivirus infection enhancement in macrophages: radioactive and biological studies on the effect of antibody on viral fate. J. Gen. Virol. **65**:1261–1272.
14. **Helenius, A., J. Kartenbeck, K. Simons, and E. Fries.** 1980. On the entry of Semliki Forest virus into BHK-21 cells. J. Cell Biol. **84**:404–420.
15. **Helenius, A., B. Morein, E. Fries, K. Simons, P. Robinson, V. Schirrmacher, C. Terhorst, and J. L. Strominger.** 1978. Human (HLA-A and HLA-B) and murine (H-2K and H-2D) histocompatibility antigens are cell surface receptors for Semliki Forest virus. Proc. Natl. Acad. Sci. USA **75**:3846–3850.
16. **Higa, H. H., G. N. Rogers, and J. C. Paulson.** 1985. Influenza virus hemagglutinins differentiate between receptor determinants bearing N-acetyl-N-glycollyl-, and N, O-diacetylneuraminic acids. Virology **144**:279–282.
17. **Holland, J. J.** 1984. Continuum of change in RNA virus genomes, p. 137–143. *In* A. L. Notkins and M. B. A. Oldstone (ed.), Concepts in viral pathogenesis. Springer-Verlag, New York.
18. **Inada, T., and C. A. Mims.** 1984. Mouse Ia antigens are receptors for lactate dehydrogenase virus. Nature (London) **309**:59–61.
19. **Klatzmann, D., E. Champagne, S. Chamaret, J. Gruest, D. Guetard, T. Hercend, J.-C. Gluckman, and L. Montagnier.** 1984. T-lymphocyte T4 molecule behaves as the receptor for human retrovirus LAV. Nature (London) **312**:767–768.
20. **Korant, B. D., K. Lonberg-Holm, and P. LaColla.** 1984. Picornaviruses and togaviruses: targets for design of antivirals, p. 61–98. *In* E. De Clerq and R. T. Walker (ed.), Targets for the design of antiviral agents. Plenum Publishing Corp., New York.
21. **Kowalchyk, K., and P. G. W. Plagemann.** 1985. Cell surface receptors for lactate dehydrogenase-elevating virus on subpopulations of macrophages. Virus Res. **2**:211–229.

22. **Landen, B., and C. F. Fox.** 1980. Isolation of BP gp 70, a fibroblast receptor for the envelope antigen of Rauscher murine leukemia virus. Proc. Natl. Acad. Sci. USA **77:**4988–4992.
23. **Lonberg-Holm, K.** 1981. Attachment of animal viruses to cells: an introduction, p. 1–20. *In* K. Lonberg-Holm and L. Philipson (ed.), Virus receptors, part 2. Chapman & Hall, Ltd., London.
24. **Lonberg-Holm, K., R. L. Crowell, and L. Philipson.** 1976. Unrelated animal viruses share receptors. Nature (London) **259:**679–681.
25. **Lonberg-Holm, K., L. B. Gosser, and E. J. Shimshick.** 1976. Interaction of liposomes with subviral particles of poliovirus type 2 and rhinovirus type 2. J. Virol. **19:**746–749.
26. **Lonberg-Holm, K., and L. Philipson.** 1974. Early interaction between animal viruses and cells. Monogr. Virol. **9:**1–148.
27. **Lonberg-Holm, K., and L. Philipson (ed.).** 1981. Virus receptors, part 2. Receptors and recognition series B, vol. 8. Chapman & Hall, Ltd., London.
28. **Lonberg-Holm, K., and N. M. Whiteley.** 1976. Physical and metabolic requirements for early interaction of poliovirus and human rhinovirus with HeLa cells. J. Virol. **19:**857–870.
29. **Mapoles, J. E., D. L. Krah, and R. L. Crowell.** 1985. Purification of a HeLa cell receptor protein for group B coxsackieviruses. J. Virol. **55:**560–566.
30. **Marchesi, V. T., H. Furthmayr, and M. Tomita.** 1976. The red cell membrane. Annu. Rev. Biochem. **45:**667–698.
31. **Markwell, M. A. K., P. Fredman, and L. Svennerholm.** 1984. Receptor ganglioside content of three hosts for Sendai virus. Biochim. Biophys. Acta **775:**1–16.
32. **Maxfield, F. R.** 1982. Weak bases and ionophores rapidly and reversibly raise the pH of endocytic vesicles in cultured mouse fibroblasts. J. Cell Biol. **95:**676–681.
33. **Miller, D. A., O. J. Miller, V. G. Dev, S. Hashmi, R. Tantravhi, L. Medrano, and H. Green.** 1974. Human chromosome 19 carries a poliovirus receptor gene. Cell **1:**167–173.
34. **Nemerow, G. R., R. Wolfert, M. E. McNaughton, and N. R. Cooper.** 1985. Identification and characterization of the Epstein-Barr virus receptor on human B lymphocytes and its relationship to the C3d complement receptor (CR2). J. Virol. **55:**347–351.
35. **Okuma, S., and B. Poole.** 1978. Fluorescence probe measurement of the intralysosomal pH in living cells and the perturbation of pH by various agents. Proc. Natl. Acad. Sci. USA **75:**3327–3331.
36. **Oldstone, M. B. A., A. Tishon, F. J. Dutko, S. I. T. Kennedy, J. J. Holland, and P. J. Lampert.** 1980. Does the major histocompatibility complex serve as specific receptor for Semliki Forest virus? J. Virol. **34:**256–265.
37. **Pastan, I. H., and M. C. Willingham.** 1981. Journey to the center of the cell: role of the receptosome. Science **214:**504–509.
38. **Racaniello, V. R.** 1984. Poliovirus type II produced from cloned cDNA is infectious in mice. Virus Res **1:**669–675.
39. **Ratner, L., R. C. Gallo, and F. Wong-Staal.** 1985. HTLV-III, LAV, ARV are variants of same AIDS virus. Nature (London) **313:**636–637.
40. **Reagan, K. J., B. Goldberg, and R. L. Crowell.** 1984. Altered receptor specificity of coxsackievirus B3 after growth in rhabdomyosarcoma cells. J. Virol. **49:**635–640.
41. **Reisert, P. S., R. C. Spiro, P. L. Townsend, S. A. Stanford, T. Sairenji, and R. E. Humphreys.** 1985. Functional association of class II antigens with cell surface binding of Epstein-Barr virus. J. Immunol. **134:**3776–3780.
42. **Schlegel, R., and M. Wade.** 1985. Biologically active peptides of the vesicular stomatitis virus glycoprotein. J. Virol **53:**319–323.
43. **Sharpe, A. H., and B. N. Fields.** 1985. Pathogenesis of viral infections. Basic concepts derived from the reovirus model. N. Engl. J. Med. **312:**486–497.
44. **Skehel, J. J., P. M. Bayley, E. B. Brown, S. R. Martin, M. D. Waterfield, J. M. White, J. A. Wilson, and C. D. Wiley.** 1982. Changes in the conformation of influenza virus hemagglutinin at the pH optimum of virus-mediated membrane fusion. Proc. Natl. Acad. Sci. USA **79:**968–972.
45. **Svensson, U., R. Persson, and E. Everitt.** 1981. Virus-receptor interactions in the adenovirus system. I. Identification of virion attachment proteins of the HeLa cell plasma membrane. J. Virol. **38:**70–81.
46. **Tardieu, M., R. L. Epstein, and H. L. Weiner.** 1982. Interaction of viruses with cell surface receptors. Int. Rev. Cytol. **80:**27–61.
47. **Tignor, G. H., A. L. Smith, and R. E. Shope.** 1984. Utilization of host proteins as virus receptors, p. 109–116. *In* A. L. Notkins and M. B. A. Oldstone (ed.), Concepts in viral pathogenesis. Springer-Verlag, New York.

48. **Tycko, B., and F. R. Maxfield.** 1982. Rapid acidification of endocytic vesicles containing alpha-2-macroglobulin. Cell **28:**643–651.
49. **White, J., and A. Helenius.** 1980. pH dependent fusion between the Semliki Forest virus membrane and liposomes. Proc. Natl. Acad. Sci. USA **77:**3273–3277.
50. **White, J., J. Kartenbeck, and A. Helenius.** 1980. Fusion of Semliki Forest virus with the plasma membrane can be induced by low pH. J. Cell Biol. **87:**264–272.
51. **White, J., K. Matlin, and A. Helenius.** 1981. Cell fusion by Semliki Forest, influenza, and vesicular stomatitis viruses. J. Cell Biol. **89:**674–679.
52. **Wunner, W. H., K. J. Reagan, and H. Koprowski.** 1984. Characterization of saturable binding sites for rabies virus. J. Virol. **50:**691–697.

Virus Attachment Proteins

Nature and Function of the Reovirus Hemagglutinin in Cell and Tissue Tropism

RHONDA BASSEL-DUBY,[1] ANULA JAYASURIYA,[1] AND
BERNARD N. FIELDS[1,2]

*Department of Microbiology and Molecular Genetics, Harvard Medical School,[1]
Shipley Institute of Medicine,[2] and Department of Medicine, Division of Infectious
Diseases, Brigham and Women's Hospital, Boston, Massachusetts 02115[2]*

Viral binding to cell surfaces is one of the major determinants of
cell and tissue tropism. The reovirus hemagglutinin is a viral
surface protein that interacts with cell surface receptors. Experiments using both viral mutants and monoclonal antibodies have
determined that one region of the hemagglutinin interacts with
receptors. Cloned cDNA copies of the S1 gene (encoding the hemagglutinin) have revealed the chemical nature of both the receptor
binding region and the coiled-coil α-helical region that anchors the
hemagglutinin protein to the viral particle. These studies provide a
molecular basis for understanding the role of particular sequences
in localizing viral infection in vivo.

VIRUS ATTACHMENT PROTEINS (VAP) AND TISSUE TROPISM

The precise properties of microorganisms that determine the stages of
pathogenesis are poorly understood. Recent studies using molecular biology
techniques have led to new insights into mechanisms of viral pathogenesis (35).
Although the infectious process is quite complex, it is clear that one critical
stage, the interaction of VAP with cell surface receptors, often plays a major role
in determining the pattern of infection (14). With increased knowledge of the
functions of viral structural components and their interaction with host cells,
molecular mechanisms of viral pathogenesis can be better defined.

Classically, infection occurs when a virus attaches itself to a host cell, gains
entry, and subsequently produces cellular injury. Distinct disease patterns are
dictated in part by the ability of different viruses to attach to different host cells.
The specificity of a virus for a particular cell is thus partially determined by the
interaction of its surface structures with receptors on the host cell. Viral tissue
tropism results from variability of the host-cell range of the viruses. Certain
viruses are highly localized in certain cell types, such as reovirus type 3 (27) and
rabies virus (28) in neurons and reovirus type 1 in ependymal cells (23). Other
viruses such as herpesvirus (30) and measles virus (3) have a more widely
distributed host-cell range.

Generally the VAP are located on the surface of the virion, accessible to
host-cell receptors. The VAP for mammalian reoviruses has been identified as
the σ1 protein (viral hemagglutinin), an outer capsid protein. For bunyaviruses
(2), rabies virus (10, 12), and paramyxoviruses and myxoviruses (7), viral
surface glycoproteins have been implicated in viral tissue tropism.

13

It is likely that the host-cell receptors for viruses are not de novo viral receptors, but existing cellular receptors with essential biological functions. There are several examples of this: Epstein-Barr virus binds to complement receptor CR2 (17); rabies virus binds to acetylcholine receptors (25); Semliki Forest virus interacts with MHC (major histocompatibility complex) antigens (21); CD4(T4) antigen has been shown to be an essential component of the receptor for the acquired immune deficiency syndrome (AIDS) retrovirus (HTLV-III) (11); lactate dehydrogenase virus has been shown to be associated with mouse Ia antigens (22); and, most recently, reovirus type 3 was shown to interact with receptors that are similar to beta-adrenergic receptors (9; discussed by Co et al., this volume).

REOVIRUS AND TISSUE TROPISM

Reovirus structure (reviewed in reference 14). The mammalian reoviruses can be subdivided into three serological subgroups: serotype 1 (strain Lang), serotype 2 (strain Jones), and serotype 3 (strain Dearing). All mammalian reoviruses consist of a segmented, double-stranded RNA genome which is surrounded by a double capsid. Both the inner capsid (core) and the outer capsid display icosahedral symmetry. The 10 segments of the viral genome make up three size classes of molecules: large (segments L1, L2, and L3), medium (segments M1, M2, and M3), and small (segments S1, S2, S3, and S4). The electrophoretic patterns of the gene segments of the three serotypes are similar but readily distinguished, and these electrophoretic migrational differences allow for parental genomic assignment of intertypic reassortants.

The double-stranded RNA genes are each transcribed into a single-stranded RNA (plus polarity) which is an identical copy of its entire gene template. The single-stranded RNA is translated into viral polypeptides which, like the genome segments that encode them, fall into three size classes: large ($\lambda 1$, $\lambda 2$, and $\lambda 3$), medium ($\mu 1 \rightarrow \mu 1c$, $\mu 2$, and μNS), and small ($\sigma 1$, $\sigma 2$, σNS, and $\sigma 3$). The viral outer capsid consists of three polypeptides, $\sigma 1$, $\sigma 3$, and $\mu 1c$, while the core consists of the viral polypeptides $\sigma 2$, $\lambda 1$, and $\lambda 2$. An obvious morphological feature of the core is 12 projections that seem to be located on 12 fivefold vertices of the icosahedron (26). These projections are believed to be five molecules of the $\lambda 2$ protein (34). The $\lambda 2$ protein projections have been demonstrated to be exposed on the surface of the virion and have also been shown to be closely associated with the $\sigma 1$ protein (VAP) (20).

Recently, the S1 gene which encodes the $\sigma 1$ protein (viral hemagglutinin) has been sequenced (1, 6, 29). Analysis of the predicted amino acid sequence of the $\sigma 1$ protein indicates that the amino-terminal portion of the protein contains an α-helical structure. A proposed model of the outer capsid is illustrated in Fig. 1. The $\sigma 1$ protein is suggested to be in a dimer form in close association with the $\lambda 2$ protein projections. The model shows the $\sigma 1$ protein dimer with its amino-terminal portion as an extended α-helical coiled-coil structure which extends into the virion through the channel of the $\lambda 2$ protein projections. The carboxy-terminal region of the $\sigma 1$ protein is shown exposed on the surface of the virion.

Reovirus $\sigma 1$ protein (viral hemagglutinin-VAP). The reovirus $\sigma 1$ protein was first implicated in reovirus-cell interactions when it was identified as the viral

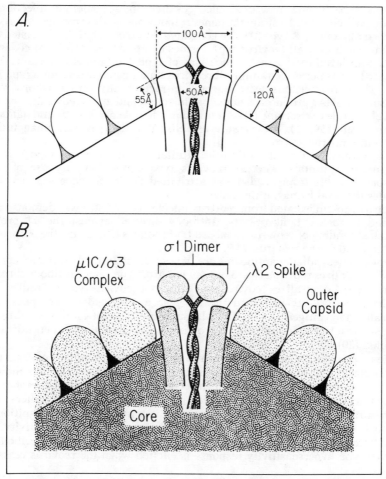

FIG. 1. Schematic representation of the morphology of the outer capsid of reovirus type 3. (A) Dimensions of the virion. (B) Orientation of the σ1 protein in the virus. The α-helical region is shown to extend through the λ2 channel into the viral core. The globular structure sits on top of the λ2 channel and interacts with host-cell receptors. (From reference 1 with permission.)

hemagglutinin (43). Reovirus hemagglutinins are type specific; type 1 aggluti-nates human erythrocytes and type 3 agglutinates bovine erythrocytes (13). Studies on reovirus tropism in the central nervous system further verify the role of the σ1 protein in virus-receptor interactions (39). Intracerebral injections of reovirus type 3 into newborn mice resulted in an infection in neurons and produced a highly lethal encephalitis (27, 33), while intracerebral injections of reovirus type 1 into newborn mice resulted in a nonlethal infection of ependymal cells (23). To study the genes responsible for this tropism, intertypic reassortants (type 1 × type 3) were generated. Reassortant 3HA1 consisted of a genome containing all reovirus type 3 genes except for the gene that codes for the hemagglutinin (S1 gene), which was of type 1 origin. Reassortant 1HA3 consisted of a genome containing all reovirus type 1 genes except for a type 3 S1 gene. These reassortants were injected intracerebrally into newborn mice. Reassortant 3HA1 produced a nonlethal infection and appeared in the ependymal cells, while reassortant 1HA3 resulted in a lethal encephalitis and damaged neuronal cells (39, 42). These results establish that the reovirus hemagglutinin determines neurotropism.

Viral tropism also exists within the pituitary gland. Reovirus type 1 infects the anterior pituitary whereas reovirus type 3 does not. By use of viral reassortants, this tropism also was attributed to the S1 gene segment that encodes the viral hemagglutinin (32).

Free σ1 protein isolated from reovirus-infected cell lysates was demonstrated to bind to mouse L fibroblasts. Antibody directed against the σ1 protein abolished binding of reovirus to mouse L fibroblasts (24). Thus, the σ1 protein is referred to as the reovirus VAP.

Reovirus-neutralizing antibodies are type specific; antibody directed against type 3 neutralizes reovirus type 3 but not reovirus type 1, and antibody directed against type 1 neutralizes reovirus type 1 but not reovirus type 3. Studies with reovirus reassortants identified the S1 gene as the major serotype-specific neutralization antigen (5, 20, 40), although monoclonal antibodies directed against the σ3 protein and the λ2 protein also elicited a weak neutralization response (20).

The σ1 protein has also been shown to be a major determinant of specificity for cellular immunity. Genetic experiments using reassortant viruses indicate that the σ1 protein is a major antigen recognized by cytotoxic T lymphocytes (16). Studies using σ1 protein neutralizing monoclonal antibodies indicate that the neutralization domain of the σ1 protein is a major site of recognition by cytotoxic T lymphocytes (15). In addition, delayed-type hypersensitivity, another T-cell-mediated immune reaction, also exhibits a serotype specificity. A delayed-type hypersensitivity response is elicited when the reovirus reassort-

---

FIG. 2. Diagrams of coronal brain sections indicating location of virus-induced lesions. (a to e) Diagrams of coronal brain sections extending from (a) rostral to (e) caudal areas of the brain, showing the location of virus-induced lesions in mice infected intracerebrally with either the Dearing strain of reovirus type 3 or a variant Dearing virus. Black areas indicate regions of the brain where Dearing virus as well as the variant viruses induced lesions; dotted areas show regions where only the Dearing virus caused necrosis. Abbreviations: ac, anterior colliculus; c, cingulum; cn, caudate nucleus; dg, dentate girus; dnh, dorsomedial nucleus of hypothalamus; h, hippocampus; lgb, lateral geniculate body; ltn, lateral thalamic nucleus; lv, lateral ventricle; mb, mammillary body; mcn, media cuneate nucleus; msn, medial septum nucleus; oc, occipital cortex; s, subiculum; sn, substantia nigra; stn, spinal trigeminal nucleus; tv, third ventricle; zi, zona incerta. (From reference 37 with permission.)

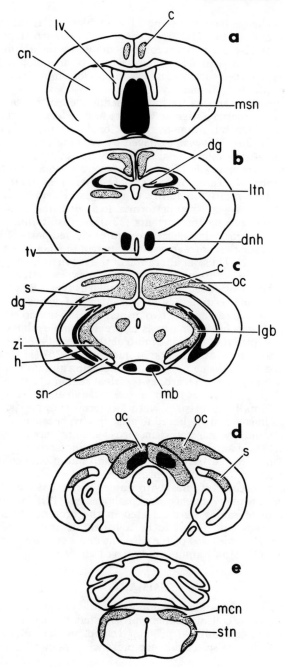

ants used for challenge have the same S1 RNA segment as the virus used for subcutaneous immunization (41). The generation of reovirus type-specific suppressor T cells has also been shown to be mediated by the σ1 protein (18, 19).

In summary, the reovirus σ1 protein, which is an outer capsid protein, interacts with various cell surfaces and is a major determinant of cell and tissue tropism and specificity in immunity.

Reovirus σ1 protein variants. To further define the neutralization site of the type 3 σ1 protein, reovirus antigenic variants were selected according to their capacity to resist neutralization by monoclonal antibodies specific for the reoviral type 3 σ1 (38). This was achieved by growing reovirus type 3 in the presence of neutralizing monoclonal antibody and isolating reovirus variants with an altered hemagglutinin. Three variants were isolated in this manner: variants A, F, and K. These variants were altered in their virulence, as determined by their relative inability to cause fatal neurological disease in newborn mice. Studies were then performed to analyze the anatomic distribution of injury produced by the variants (37). The results indicated that the variants were altered in their capacity to injure selected parts of the brain and were unable to invade the central nervous system of the mouse when inoculated intracerebrally (Fig. 2). Thus, an alteration of the reovirus hemagglutinin at its neutralization site altered tissue tropism in the nervous system.

Comparison of nucleotide sequence of the reovirus type 3 S1 gene and variant K S1 gene. The reovirus type 3 and variant K S1 genes each have 1,416 nucleotide bases which can encode two proteins: a 120-amino acid protein and a 455-amino acid protein. On the basis of previous molecular weight estimations, the hemagglutinin is the 455-amino acid protein (4, 36). Comparison of the sequences shows a point mutation at nucleotide 1,267 (1). This results in a change at residue 419 in the hemagglutinin, replacing glutamic acid with lysine and altering the protein at the carboxy-terminal region.

The discovery that a reovirus immune-selected avirulent variant has a point mutation is not totally surprising. Naturally occurring persistent infections with visna virus in sheep are believed to arise in vivo because of antigenic drift (31). Cell culture immune-selected visna virus variants were used to study viral genetic mechanisms of antigenic variation. Results indicated that the molecular basis for the visna virus antigenic drift was point mutational changes in the 3' region of the gene for the viral coat protein (8).

In addition, molecular studies with rabies virus have also shown that antigenic alterations may influence viral pathogenesis (12). Variants of a highly pathogenic strain of rabies were selected in the presence of a neutralizing monoclonal antibody and found to be avirulent in mice. Sequencing data obtained for the glycoproteins of the virulent and avirulent variants showed that substitution of a single amino acid altered virulence.

SUMMARY

Cell attachment proteins play an important role in determining viral tissue tropism. The reovirus cell attachment protein has been identified as the σ1 protein (viral hemagglutinin), a protein located on the outer capsid of the virion. Reovirus variants selected in the presence of neutralizing antibody are avirulent in newborn mice and show altered tropism. Sequencing data obtained for the cell attachment protein of reovirus type 3 and variant K show that a single amino acid substitution alters tissue tropism.

We thank Kenneth Tyler and Max Nibert for insightful discussions, Edmund Choi and Jonathan Seidman for help with DNA sequencing and useful discussions, and Marcia Masters for typing this manuscript.

This work was supported by Public Health Service research grant 5 RO1 AI 13178 from the National Institute of Allergy and Infectious Diseases and by Shipley Institute of Medicine.

LITERATURE CITED

1. **Bassel-Duby, R., A. Jayasuria, D. Chatterjee, N. Sonenberg, J. V. Maizel, and B. N. Fields.** 1985. Sequence of reovirus haemagglutinin predicts a coiled-coil structure. Nature (London) **315:**421–423.
2. **Beatty, B. J., B. R. Miller, R. E. Shope, E. J. Rozhon, and H. L. Bishop.** 1982. Molecular basis of bunyavirus infection of mosquitoes: role of the middle-sized RNA segment. Proc. Natl. Acad. Sci. USA **79:**1295.
3. **Black, F. L.** 1984. Measles, p. 397–414. *In* A. S. Evans (ed.), Viral infections of humans, epidemiology and control. Plenum Publishing Corp., New York.
4. **Both, G. W., S. Lavi, and A. J. Shatkin.** 1975. Synthesis of all the gene products of the reovirus genome *in vivo* and *in vitro*. Cell **4:**173–180.
5. **Burstin, S. J., D. R. Spriggs, and B. N. Fields.** 1982. Evidence for functional domains on the reovirus type 3 hemagglutinin. Virology **117:**146–155.
6. **Cashdollar, L. W., R. A. Chmelo, J. R. Weiner, and W. K. Joklik.** 1985. The sequence of the S1 genes of the three serotypes of reovirus. Proc. Natl. Acad. Sci. USA **82:**24–28.
7. **Choppin, P. W., and A. Scheid.** 1980. The role of viral glycoproteins in adsorption, penetration and pathogenicity of viruses. Rev. Infect. Dis. **2:**40–61.
8. **Clements, J. E., N. D'Antonio, and O. Narayan.** 1982. Genomic changes associated with antigenic variation of visna virus. J. Mol. Biol. **158:**415–434.
9. **Co, M. S., G. N. Gaulton, A. Tominaga, C. J. Homcy, B. N. Fields, and M. L. Greene.** 1985. Structural similarities between the mammalian beta-adrenergic and reovirus type 3 receptors. Proc. Natl. Acad. Sci. USA **82:**5315–5318.
10. **Coulon, P., P. Rollin, M. Aubert, and A. Flamand.** 1982. Molecular basis of rabies virus virulence. I. Selection of avirulent mutants of the CVS strain with anti-G monoclonal antibodies. J. Gen. Virol. **61:**97–100.
11. **Dalgleish, A. G., P. C. L. Beverley, P. R. Clapham, D. H. Crawford, M. F. Greaves, and R. A. Weiss.** 1984. The CD4 (T4) antigen is an essential component of the receptor for the AIDS retrovirus. Nature (London) **312:**763–767.
12. **Dietzschold, G., W. H. Wunner, T. J. Wiktor, A. D. Lopes, M. Lafon, C. L. Smith, and H. Koprowski.** 1983. Characterization of an antigenic determinant of the glycoprotein that correlates with pathogenicity of rabies virus. Proc. Natl. Acad. Sci. USA **80:**70–74.
13. **Eggers, H. J., P. J. Gomatos, and I. Tamm.** 1962. Agglutination of bovine erythrocytes: a general characteristic of reovirus type 3. Proc. Soc. Exp. Biol. Med. **110:**879–881.
14. **Fields, B. N., and M. I. Greene.** 1982. Genetic and molecular mechanisms of viral pathogenesis: implications for prevention and treatment. Nature (London) **300:**19–23.
15. **Finberg, R., D. R. Spriggs, and B. N. Fields.** 1982. Host immune response to reovirus: CTL recognize the major neutralization domain of the viral hemagglutinin. J. Immunol. **129:**2235–2238.
16. **Finberg, R., H. L. Weiner, B. N. Fields, B. Benacerraf, and S. J. Burakoff.** 1979. Generation of cytolytic T lymphocytes after reovirus infection: role of S1 gene. Proc. Natl. Acad. Sci. USA **76:**442–446.
17. **Fingerorth, J. D., J. J. Weis, T. F. Tedder, J. L. Stominger, P. A. Psiro, and D. T. Fearon.** 1984. Epstein-Barr virus receptor of human B lymphocytes as the C3d receptor CR2. Proc. Natl. Acad. Sci. USA **81:**4510–4514.
18. **Fontana, A., and H. L. Weiner.** 1980. Interaction of reovirus with cell surface receptors. II. Generation of suppressor T cells by the hemagglutinin of reovirus type 3. J. Immunol. **125:**2660–2664.
19. **Greene, M. I., and H. L. Weiner.** 1980. Delayed hypersensitivity in mice infected with reovirus. II.

Induction of tolerance and suppressor T cells to viral specific gene products. J. Immunol. **125**:283–287.

20. **Hayes, E. C., P. W. K. Lee, S. E. Miller, and W. K. Joklik.** 1981. The interaction of a series of hybridoma IgGs with reovirus particles: demonstration that the core protein λ2 is exposed on the particle surface. Virology **108**:147–155.
21. **Helenius, A., B. Morein, E. Fries, K. Simons, P. Robinson, V. Schirrmacher, C. Terhorst, and J. Strominger.** 1978. Human (HLA-A and HLA-B) and murine (H-2K and H-2D) histocompatibility antigens are cell surface receptors for Semliki Forest virus. Proc. Natl. Acad. Sci. USA **75**:3846–3850.
22. **Inada, T., and C. A. Mims.** 1984. Mouse Ia antigens are receptors for lactate dehydrogenase virus. Nature (London) **309**:59–61.
23. **Kilham, L., and G. Margolis.** 1969. Hydrocephalus in hamsters, ferrets, rats and mice following inoculations with reovirus type 1. Lab. Invest. **21**:183–188.
24. **Lee, P. W. K., E. C. Hayes, and W. K. Joklik.** 1981. Protein σ1 is the reovirus cell attachment protein. Virology **108**:156–163.
25. **Lentz, T. L., T. G. Burrage, A. L. Smith, J. Crick, and G. H. Tignor.** 1982. Is the acetylcholine receptor a rabies virus receptor? Science **215**:182–184.
26. **Luftig, R. B., S. Kilham, A. J. Hay, H. J. Zweerink, and W. K. Joklik.** 1972. An ultrastructure study of virions and cores of reovirus type 3. Virology **48**:170–181.
27. **Margolis, G., L. Kilham, and N. Gonatos.** 1971. Reovirus type III encephalitis: observations of virus-cell interactions in neural tissues. I. Light microscopy studies. Lab. Invest. **24**:91–100.
28. **Murphy, F. A.** 1977. Rabies pathogenesis. Arch. Virol. **54**:279–297.
29. **Nagata, L., S. A. Masri, C. W. Mah, and P. W. K. Lee.** 1984. Molecular cloning and sequencing of the reovirus (serotype 3) S1 gene which encodes the viral cell attachment protein σ1. Nucleic Acids Res. **12**:8699–8710.
30. **Nahmias, A. J., and W. E. Josey.** 1984. Herpes simplex viruses 1 and 2, p. 351–372. In A. S. Evans (ed.), Viral infections of humans, epidemiology and control. Plenum Publishing Corp., New York.
31. **Narayan, O., D. E. Griffin, and J. Chase.** 1977. Antigenic shift of visna virus in persistently infected sheep. Science **197**:376–378.
32. **Onodera, T., A. Toniolo, U. R. Ray, A. B. Jensen, R. A. Knazek, and A. L. Notkins.** 1981. Virus-induced diabetes mellitus. XX. Polyendocrinopathy and autoimmunity. J. Exp. Med. **153**:1457–1473.
33. **Raine, C. S., and B. N. Fields.** 1973. Ultrastructural features of reovirus type 3 encephalitis. J. Neuropathol. Exp. Neurol. **32**:19–33.
34. **Ralph, S. J., J. D. Harvey, and A. R. Bellamy.** 1980. Subunit structure of the reovirus spike. J. Virol. **36**:894.
35. **Smith, H., J. J. Skehel, and M. J. Turner (ed.).** 1980. The molecular basis of microbial pathogenicity. Verlag-Chemie, Basel.
36. **Smith, R. E., H. J. Zweerink, and W. K. Joklik.** 1969. Polypeptide components of virions, top component and cores of reovirus 3. Virology **39**:791–810.
37. **Spriggs, D. R., R. T. Bronson, and B. N. Fields.** 1983. Haemagglutinin variants of reovirus type 3 have altered central nervous system tropism. Science **220**:505–507.
38. **Spriggs, D. R., and B. N. Fields.** 1982. Attenuated reovirus type 3 strains generated by selection of hemagglutinin antigenic variants. Nature (London) **297**:68–70.
39. **Weiner, H. L., D. Drayna, D. R. Averill, Jr., and B. N. Fields.** 1977. Molecular basis of reovirus virulence: role of the S1 gene. Proc. Natl. Acad. Sci. USA **74**:5744–5748.
40. **Weiner, H. L., and B. N. Fields.** 1977. Neutralization of reovirus: the gene responsible for the neutralization antigen. J. Exp. Med. **146**:1305–1310.
41. **Weiner, H. L., M. I. Greene, and B. N. Fields.** 1980. Delayed hypersensitivity in mice infected with reovirus. I. Identification of host and viral gene products responsible for the immune response. J. Immunol. **125**:278–282.
42. **Weiner, H. L., M. L. Powers, and B. N. Fields.** 1980. Absolute linkage of virulence and central nervous system tropism of reoviruses to viral hemagglutinin. J. Infect. Dis. **141**:609–616.
43. **Weiner, H. L., R. F. Ramig, T. A. Mustoe, and B. N. Fields.** 1978. Identification of the gene coding for the hemagglutinin of reovirus. Virology **86**:581–584.

Location of Four Neutralization Antigens on the Three-Dimensional Surface of a Common-Cold Picornavirus, Human Rhinovirus 14

ROLAND RUECKERT,[1] BARBARA SHERRY,[1] ANNE MOSSER,[1] RICHARD COLONNO,[2] AND MICHAEL ROSSMANN[3]

Biophysics Laboratory and Department of Biochemistry, University of Wisconsin, Madison, Wisconsin 53706[1]; Merck, Sharp and Dohme Research Laboratories, West Point, Pennsylvania 19486[2]; and Department of Biological Sciences, Purdue University, West Lafayette, Indiana 47907[3]

A panel of neutralizing murine monoclonal antibodies were raised against intact human rhinovirus 14, and 62 neutralization-resistant virus mutants were selected with these. Cross-neutralization tests resolved the mutants into four neutralization immunogen groups. The mutations were clustered in the serotype-variable regions of the amino acid sequences of the virus capsid proteins. X-ray crystallographic analysis of human rhinovirus 14 shows the protomer surface to contain a 25 Å (2.5 nm) deep cleft which is probably the binding site for host-cell receptors. One side of this cleft is formed by VP1; the other side, by VP2 and VP3. Mutations which led to resistance to monoclonal antibodies occurred without exception in amino acids which protrude from the protomer surface in four well-defined clusters. These are the proposed neutralization immunogen antibody binding sites. The floor of the cleft appears to be protected from attack by antibody molecules, the diameters of which are too great for penetration.

The picornaviruses are a family of nonenveloped particles about 30 nm in diameter (7). Each particle contains a single molecule of infective RNA packaged within the central cavity of a protein shell about 5 nm thick. Numbered among the picornaviruses are poliovirus, human hepatitis A virus, foot-and-mouth disease virus, and the human rhinoviruses. A large backlog of research on these medically and agriculturally important viruses has produced a library of over 230 serotypes and reference antisera. These reagents constitute an invaluable resource, not only for clinical diagnosis of picornaviral disease agents but also for studies on the molecular basis of immunogenicity and serotypic variation.

All picornaviruses contain four polypeptide chains (Fig. 1). The hypothesis originally put forth in 1969 (8), that these chains are organized into 60 protomers arranged into 12 pentamers (5, 8, 15), has now been generally accepted (4, 10), and attention has since focused upon mapping topological relationships of the four chains within the capsid by using methods such as chemical labeling of the surface of intact particles, treatment with cross-linking reagents, reaction with specific antibodies, and cross-linking with UV light

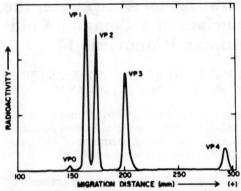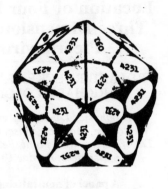

FIG. 1. Dodecahedral model of picornavirus capsid structure (5, 11). The four peaks, separated from dissociated virions by electrophoresis on a sodium dodecyl sulfate-polyacrylamide gel (left), are derived from identical four-segmented protein subunits called "mature" protomers (VP4, VP2, VP3, VP1). The protomer is defined as the smallest identical subunit of an oligomeric protein. Traces of protein VP0, an uncleaved precursor of VP2 and VP4, reflect the presence of "immature" protomers (VP0, VP3, VP1) in the virion shell (see periphery at one o'clock, right). The 60 protomers are organized as 12 pentamers (bounded by dotted lines).

(reviewed in reference 10). The consensus is that VP4 lies in the interior near the RNA while the three larger proteins are exposed, with VP1 being the most dominant surface protein.

Identification of immunogenic neutralization sites will be useful for studies on the mechanism of neutralization, as a starting point for developing subunit vaccines and as a foundation for understanding the molecular basis of serotypic variation. Type 1 poliovirus and type 14 human rhinovirus (HRV-14) make particularly interesting objects for comparison in this regard. Their genomes have been sequenced and have revealed an unexpectedly close relationship in the sequence of their coat proteins (3, 14). Yet the two viruses occupy different biological niches, use different receptors (1), and differ markedly in serotype variability (>89 serotypes in human rhinoviruses compared with only 3 in the polioviruses).

We and others have set out to map the location of polio- and rhinoviral antigens with the aid of neutralizing monoclonal antibodies raised against native virions. Such antibodies identify the most highly immunogenic neutralization antigens, i.e., those most likely to confer immune protection after vaccination or natural infection.

To study the antigenic structure of HRV-14, we prepared a panel of 35 neutralizing murine monoclonal antibodies, raised against intact virus, and used some of these antibodies to select neutralization-resistant mutants, 62 in all (12, 13). Cross-neutralization against the panel of antibodies resolved the mutants into four operational groups (IA, IB, II, and III) called neutralization immunogens (NIm) (Fig. 2).

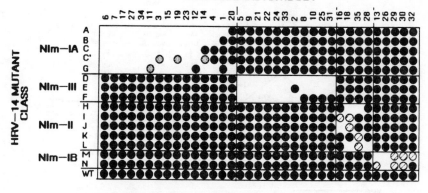

FIG. 2. Classification of each monoclonal antibody into one of four groups (IA, IB, II, and III) by neutralization pattern with a panel of resistant mutant viruses.

The rhinovirus mutants were found, by isoelectric focusing, to be charge altered in VP1 (IA and IB), VP2 (II), or VP3 (III) (13). Sequencing by the dideoxy method revealed mutated amino acids in a relatively limited number of sites. As shown in Fig. 3, most of these mutations were clustered in serotype-variable regions of the amino acid sequence, but four were far removed in the sequence from their sister clusters.

Development of methods for purification and crystallization of poliovirus and HRV-14 has opened the way for structural analysis by the methods of X-ray crystallography. We chose HRV-14 for its stability and reasonably good yields (about 250 μg per 10^9 cells) in culture and because safety concerns were less serious than with poliovirus. After some preliminary difficulties with radiation sensitivity, conditions were found for producing crystals of satisfactory stability in the X-ray beam (2, 6). The low safety hazard of the virus alleviated problems with transport of the crystals to the Cornell high-energy facility where synchrotron radiation, which caused significantly less crystal damage than conventional X-ray sources, afforded excellent diffraction patterns. Another innovation was use of the molecular replacement technique to solve the phase problem (9). The success of the molecular replacement technique, which depends upon the prodigious speed of a supercomputer, opens the way for a rapid series of structure determinations on other picornaviruses because dependence on the riskiest step, production of isomorphous heavy atom derivatives, is now greatly reduced.

As shown in Fig. 4, the surface of the protomer is dominated by a large cleft, some 2.5 nm deep and 1.2 to 3.0 nm wide. The cleft, probably the locus of the

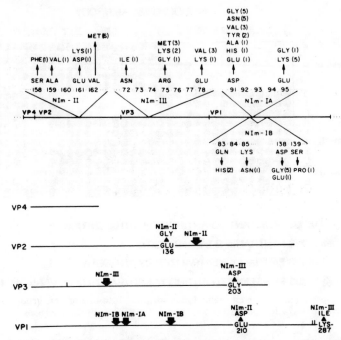

FIG. 3. Amino acid substitution sites in neutralization-resistant mutant viruses (12). All mutants resistant to NIm-IA antibodies were substituted at position 91 (18 cases) or 95 (6 cases). Those resistant to NIm-IB were mutated at VP1 residues 83 and 85 or 138 and 139 flanking either side of the IA mutations. Mutated sites in NIm-II were clustered at positions 158, 159, 161, and 162 in VP2 with one outlier" at residue 136; a second outlier was found at amino acid 210 in VP1. NIm-III mutations were similarly arranged in a cluster on VP3, (72, 75, and 78) with one outlier at position 203 and a second outlier on VP1 at position 287, just two residues from the carboxy-terminal tyrosine.

host-cell receptor-binding site, is formed by plateaus rising from the shell-forming domains of VP1 on one side and from VP2 and VP3 on the other. VP1 is the dominant surface protein forming, in addition to one of the plateaus, most of the canyon walls and floor (not visible in the 1.0-nm-thick section); VP1 also invades parts of the VP3 plateau and abuts VP2 at the canyon wall. The small polypeptide, VP4, is buried spanning the bases of VP1 to VP3 which face the inner cavity of the shell (not visible).

The absence of neutralization resistance mutations in VP4 is consistent with its internal location where antibodies are presumably unable to reach. The mutable amino acids illustrated in Fig. 3 were found, without exception, to protrude at the surface of the protomer in well-defined clusters (Fig. 4, bottom). Thus IA and IB mutations occupied separate clusters on the surface of VP1; NIm-II mutations clustered on the surface of VP2; and NIm-III mutations grouped on the surface of VP3. Three of the NIm sites (IB, II, and III) were

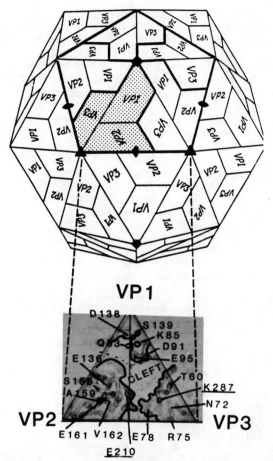

VP1

VP2

VP3

D 138
S 139
K 85
Q 93
D 91
E 136
E 95
CLEFT
S 158
T 60
A 159
K 287
N 72
E 161 V 162 E 78 R 75
E 210

FIG. 4. (Bottom) Surface features of the repeating crystallographic structure unit at 0.5-nm resolution. The contours correspond to electron density of a section cut tangentially through the subunit at a radius of 14 to 15 nm to produce a slice of the surface 1.0 nm thick. The cleft (traced with heavy lines) is devoid of electron density because its floor lies at a radius of about 12.5 nm. Pointers identify the plateau positions of the mutable amino acids (see Fig. 3). Underlined residues (K287 in NIm-III and E210 in NIm-II) represent side chains contributed by VP1. A, Alanine; D, aspartate; E, glutamate; K, lysine; N, asparagine; Q, glutamine; S, serine; T, threonine; and V, valine. (Top) Relation of one of the protomers (stippled) in the virion to the crystallographic structure unit, illustrating that the two are not identical. As a result the positions of VP2 and VP3 in the promoter are inverted relative to positions shown in the crystallographic subunit.

noncontiguous, i.e., composed of more than one segment of polypeptide backbone. NIm-II and NIm-III contained a segment of VP1. Moreover, side chains of the mutable residues always pointed outward, while those of the protruding unchanged residues usually did not. We propose that the protruding clusters defined by the mutable amino acids lie within the actual antibody binding site. There was no evidence of mutations which altered antigenic structure over a long range; this conclusion is similar to that drawn by Webster and co-workers (16), who mapped neutralization antigens on the hemagglutinin protein of influenza, an enveloped virus. These results encourage us to propose that the monoclonal mapping procedure will be useful for mapping neutralization antigens of the nonenveloped as well as the enveloped viruses.

Finally, it may be worth noting that the plateaus of VP1, VP2, and VP3 probably protect the floor of the cleft from attack by antibody molecules whose 3.5-nm diameter is too large for deep penetration into the 1.2- to 3.0-nm opening. Thus the plateaus might represent a device evolved by picornaviruses to allow variation of nonessential surface features while preserving the essential receptor-recognition sites. Thus, if the cleft is indeed the locus of the receptor-recognizing site, it may never be possible to obtain anti-idiotypic antibodies against picornavirus receptors.

LITERATURE CITED

1. **Abraham, G., and R. J. Colonno.** 1984. Many rhinovirus serotypes share the same cellular receptor. J. Virol. **51:**340–345.
2. **Arnold, E., J. W. Erickson, G. S. Fout, E. A. Frankenberger, H.-J. Hecht, M. Luo, M. G. Rossmann, and R. R. Rueckert.** 1984. Virion orientation in cubic crystals of the human common cold virus, HRV-14. J. Mol. Biol. **177:**417–430.
3. **Callahan, P., S. Mizutani, and R. J. Colonno.** 1985. Molecular cloning and complete sequence determination of the RNA genome of human rhinovirus type 14. Proc. Natl. Acad. Sci. USA **82:**732–736.
4. **Crowell, R. R. L., and B. J. Landau.** 1983. Receptors in the initiation of picornavirus infections, p. 1–42. *In* H. Fraenkel-Conrat and R. R. Wagner (ed.), Comprehensive virology, vol. 18. Plenum Publishing Corp., New York.
5. **Dunker, A. K., and R. R. Rueckert.** 1971. Fragments generated by pH dissociation of ME virus and their relation to the structure of the virion. J. Mol. Biol. **58:**217–235.
6. **Erickson, J. W., E. A. Frankenberger, M. G. Rossmann, G. S. Fout, K. C. Medappa, and R. R. Rueckert.** 1983. Crystallization of a common cold virus, human rhinovirus 14: isomorphism with poliovirus crystals. Proc. Natl. Acad. Sci. USA **80:**931–934.
7. **Matthews, R. E. F.** 1982. Classification and nomenclature of viruses. Intervirology **17:**1–199.
8. **Medappa, K. C., C. McLean, and R. R. Rueckert.** 1971. The structure of human rhinovirus 1A. Virology **44:**259–270.
9. **Rossmann, M. G., E. Arnold, J. W. Erickson, E. A. Frankenberger, J. P. Griffith, H.-J. Hecht, J. E. Johnson, G. Kamer, M. Luo, A. G. Mosser, R. R. Rueckert, B. Sherry, and G. Vriend.** 1985. Structure of a human common cold virus (rhinovirus 14) and functional relationship to other picornaviruses. Nature (London) **317:**145–153.
10. **Rueckert, R.** 1985. Picornaviruses and their replication, p. 705–738. *In* B. N. Fields (ed.), Virology. Raven Press, New York.
11. **Rueckert, R. R., A. K. Dunker, and C. M. Stoltzfus.** 1969. The structure of Mause-Elberfeld virus: a model. Proc. Natl. Acad. Sci. USA **62:**912–919.
12. **Sherry, B., A. G. Mosser, R. J. Colonno, and R. R. Rueckert.** 1985. Use of monoclonal antibodies to identify four neutralization immunogens on a common cold picornavirus, human rhinovirus 14. J. Virol. **57:**246–257.
13. **Sherry, B., and R. Rueckert.** 1985. Evidence for at least two dominant neutralization antigens on human rhinovirus 14. J. Virol. **53:**137–143.
14. **Stanway, G., P. J. Hughes, R. C. Mountford, P. D. Minor, and J. W. Almond.** 1984. The complete nucleotide sequence of a common cold virus: human rhinovirus 14. Nucleic Acids Res. **12:**7859–7875.

15. **Stoltzfus, C. M., and R. R. Rueckert.** 1972. Capsid polypeptides of Maus-Elberfeld virus. J. Virol. **10:**347–408.
16. **Webster, R. G., W. G. Laver, and G. M. Air.** 1983. Antigenic variation among type A influenza viruses, p. 127–168. *In* P. Palese and D. W. Kingsbury (ed.), Genetics of influenza viruses. Springer-Verlag, New York.

Monoclonal Antibodies as Probes of the Poliovirus Surface and of the Cellular Receptor

P. D. MINOR, D. EVANS, M. FERGUSON, J. W. ALMOND, AND
G. C. SCHILD

*National Institute for Biological Standards and Control, Hampstead, London
NW1 6RB, United Kingdom*

Antigenic variants of all three serotypes of poliovirus have been isolated and characterized by the use of a large collection of monoclonal antibodies with a view to identifying important sites in the antigenic structure of the virus. The mutations have been shown to be clustered into distinct loci, but the frequency with which the different loci are detected varies between the serotypes. These findings, and the properties of monoclonal antibodies specific for the cellular receptor for poliovirus, are reviewed.

The polioviruses are members of the family *Picornaviridae*, the virions of which contain an RNA genome of messenger sense, approximately 7,400 nucleotides in length, enclosed by a capsid made up of 60 copies each of the four virion proteins VP1, VP2, VP3, and VP4. There are three distinct serotypes of poliovirus, and the nucleotide sequence of the genome RNA is known for at least one strain of each serotype (8, 12, 13). X-ray crystallographic studies of the type 1 poliovirus strain Mahoney are in progress (7). The initial events of infection, which include attachment, penetration, and uncoating of the virus, are sensitive to neutralizing antibodies. We have therefore sought to identify the regions of the virion to which neutralizing monoclonal antibodies bind, on the hypothesis that such regions are likely to be either close to or coincident with the areas of the viral surface involved in the early events of infection. We have also successfully obtained monoclonal antibodies to the cellular receptor for poliovirus.

ANTIGENIC STRUCTURE OF TYPE 3 POLIOVIRUS

Several different preparations of monoclonal antibodies specific for type 3 poliovirus have been made over a period of time, six from BALB/c mice and two from Lewis rats. Fifty-eight monoclonal antibody preparations of various immunoglobulin classes and specificities for full and empty capsids have now been obtained, of which 27 have neutralizing activity. None of the neutralizing antibodies reacted with separated viral proteins in a Western blot analysis, although six of the non-neutralizing antibodies, which reacted only with empty capsids, did so. The epitopes responsible for neutralization of virus infectivity therefore appear to be dependent on the conformation of the proteins in the virions (5). We have consequently used indirect methods to identify the antigenically important areas of the virus, by selecting mutants no longer susceptible to antibodies which are able to neutralize the parental virus. It was assumed that most mutations in such cases were within the antibody binding region of the virion.

Mutants were isolated by two cycles of plaque purification in the presence of antibody (11) at frequencies of one mutant per 10^3 to 10^4 PFU of parental virus for the type 3 strain P3Leon/USA/1937 and at frequencies of one per 10^4 to 10^5 PFU of parental virus for the Sabin type 3 vaccine strain P3Leon $12a_1b$. Sixteen antigenically distinct mutants were identified from a total of 200 plaques picked from P3Leon/USA/1937. The pattern of reaction of the mutants in neutralization tests strongly implied that all antibodies capable of neutralizing P3Leon/USA/1937 were directed against the same antigenic site, as all mutants were resistant to at least two and usually a majority of the antibodies tested. This site is designated site 1.

Sixteen antigenically distinct mutants were identified from 103 plaques picked from the Sabin type 3 vaccine strain P3Leon $12a_1b$. In this case, however, 15 of the mutants were resistant to antibodies specific for site 1 of P3Leon/USA/1937 but were still sensitive to a monoclonal antibody designated 138, which was extremely specific for the Sabin type 3 vaccine strain (6). Conversely, the sixteenth mutant was resistant to antibody 138, but still susceptible to all other antibodies of the panel. Antibody 138 therefore appears to be directed against a separate antigenic site which could be mutated without affecting the reactions of site 1. This separate site is designated site 2.

The preliminary localization of site 1 mutations was aided by sodium dodecyl sulfate-polyacrylamide gel electrophoresis of the polypeptides of extracts of infected cells, which revealed differences in VP1 for some mutants, and by the fortuitous discovery that a particular T1-resistant oligonucleotide derived from the VP1 portion of the genome of some mutants showed characteristic alterations on two-dimensional gel electrophoresis. The site of the mutation was thus tentatively identified as being approximately 300 nucleotides from the start of the region coding for VP1 (11). Primer extension sequencing of virion RNA has now identified all the mutations for P3Leon/USA/1937 and all but one from P3Leon $12a_1b$ within site 1 (4; P. D. Minor, D. M. A. Evans, M. Ferguson, G. C. Schild, G. D. Westrop, and J. W. Almond, J. Gen. Virol., in press).

The locations of the predicted amino acid substitutions within VP1 for site 1 mutants are shown in Fig. 1 as filled squares, each square representing a different substitution. All but two were found to be within a region 89 to 100 amino acids from the N terminus of VP1. Two mutants of P3Leon $12a_1b$ had amino acid substitutions away from this region, at amino acids 166 and 253, respectively, and are believed to affect the conformation of the principal cluster. Where mutants of P3Leon/USA/1937 and P3Leon $12a_1b$ had the same amino substitution, they also had identical patterns of antigenicity.

The location of the site involved in the neutralization of the Sabin strain by antibody 138 (site 2) was made possible by the fact that the structural portion of the genome of the non-neutralized strain P3Leon/USA/1937 differs by only two amino acids from that of the sensitive Sabin strain P3Leon $12a_1b$ (12). The substitutions are in VP1 at residue 286 and in VP3 at residue 83. A recombinant prepared by G. D. Westrop having the VP1 sequence of Leon and the VP3 sequence of Sabin failed to react with antibody 138, thus implicating residue 286 in VP1. A mutant resistant to antibody 138 had a substitution at position

P3/Leon/USA/37 - VP1

```
G I E D L I S E V A Q G A L T L S L P K Q Q D S L P D T K A S G P A H S K E V P A L T A V E T G A T
                                                                                                  50

N P L A P S D T V Q T R H V V Q R R S R S E S T I E S F F A R G A C V A I I E V D N E Q P T T R A Q
                                                                                                  100

K L F A M W R I T Y K D T V Q L R R K L E F F T Y S R F D M E F T F V V T A N F T N A N N G H A L N
                                                                                                  150

Q V Y Q I M Y I P P G A P T P K S W D D Y T W Q T S S N P S I F Y T Y G A A P A R I S V P Y V G L A
                                                                                                  200

N A Y S H F Y D G F A K V P L K T D A N D Q I G D S L Y S A M T V D D F G V L A V R V V N D H N P T
                                                                                                  250

K V T S K V R I Y M K P K H V R V W C P R P P R A V P Y Y G P G V D Y K N N L D P L S E K G L T T Y
                                                                                                  300
```

P3/Leon 12a₁b - VP1

```
G I E D L I S E V A Q G A L T L S L P K Q Q D S L P D T K A S G P A H S K E V P A L T A V E T G A T
                                                                                                  50

N P L A P S D T V Q T R H V V Q R R S R S E S T I E S F F A R G A C V A I I E V D N E Q P T T R A Q
                                                                                                  100

K L F A M W R I T Y K D T V Q L R R K L E F F T Y S R F D M E F T F V V T A N F T N A N N G H A L N
                                                                                                  150

Q V Y Q I M Y I P P G A P T P K S W D D Y T W Q T S S N P S I F Y T Y G A A P A R I S V P Y V G L A
                                                                                                  200

N A Y S H F Y D G F A K V P L K T D A N D Q I G D S L Y S A M T V D D F G V L A V R V V N D H N P T
                                                                                                  250

K V T S K V R I Y M K P K H V R V W C P R P P R A V P Y Y G P G V D Y R N N L D P L S E K G L T T Y
                                                                                                  300
```

FIG. 1. Amino acid sequence of VP1 of P3Leon/USA/1937 and the Sabin type 3 poliovirus vaccine strain P3Leon 12a₁b showing the location of amino acid substitutions in mutants resistant to monoclonal antibodies. Symbols: ■, substitutions for mutants resistant to site 1 antibodies; □, silent mutation; ♦, substitutions for variants resistant to site 2 antibodies.

287, and a series of isolates of type 3 poliovirus from an immune deficient patient infected with the type 3 Sabin vaccine all failed to react with 138 and had a substitution at position 288. Site 2 is therefore considered to include positions 286 to 288 of VP1 of the Sabin type 3 vaccine strain, indicated by the filled diamonds in Fig. 1.

For both site 1 and site 2, resistance of a strain to an antibody was associated with a failure to bind the antibody (6).

Of the 27 monoclonal antibodies with neutralizing activity for type 3 poliovirus, 25 were shown to be affected by mutations within site 1 and 1 was shown to be affected by mutations within site 2. The remaining antibody was extremely specific for the Saukett strain of poliovirus type 3 and has not been studied further. Site 1 is thus strongly immunodominant in type 3 poliovirus.

Since site 1 represents the portion of the virus to which the majority of neutralizing antibodies bind, it was possible to induce neutralizing antibodies with appropriate peptides containing the amino acid sequence of VP1 between residues 89 and 100 (M. Ferguson, D. M. A. Evans, D. I. Magrath, P. D. Minor, J. W. Almond, and G. C. Schild, Virology, in press). Rabbits were immunized with a conjugated peptide having the amino acid sequence of VP1 residues 89 to 104, with additional cysteine residues at the N and C termini. The resulting sera consistently neutralized a broad range of type 3 poliovirus isolates but did not neutralize type 1 or type 2 strains. Single amino acid substitutions within site 1 rendered the virus resistant to neutralization by antipeptide sera, in a way similar to monoclonal antibodies.

Attempts were made to demonstrate binding of the monoclonal antibodies to the site 1 peptide by enzyme-linked immunosorbent assay (ELISA). However, this method gave both false-negative and false-positive results. For example, a monoclonal antibody produced by immunizing mice with the site 1 peptide (M. Ferguson, unpublished data) reacted with purified virus, but poorly if at all with the peptide in an ELISA. Negative findings in such a test therefore do not imply that the amino acid sequence recognized by the antibody on the native antigen is absent from the peptide. This conclusion is in agreement with the findings of other workers (3). In addition to this, 2 of 10 monoclonal antibodies specific for influenza hemagglutinin reacted with the site 1 poliovirus peptide as well as or better than many of the type 3 specific antibodies, although there was no sequence homology between the peptide and the hemagglutinin molecule in question. Thus in our hands, positive reactions in an ELISA do not necessarily indicate that the sequence recognized by the antibody on the native antigen is present in the peptide, and we have not employed this technique further as a means of identifying the sequences to which monoclonal antibodies bind.

Reports of other workers (1, 2) have suggested the presence of multiple immunodominant antigenic sites in the neutralization of poliovirus type 1, in contrast to our findings with type 3. It was possible that such differences could have arisen as a result of different methods of preparing the antibodies and mutants, and we have consequently performed studies of the antigenic structures of both type 2 and type 1 using the same methods that were applied to poliovirus type 3.

FIG. 2. Amino acid sequence of VP1 of the Sabin type 2 poliovirus vaccine strain showing the location of amino acid substitutions in mutants resistant to monoclonal antibodies (■).

ANTIGENIC STRUCTURE OF TYPE 2 POLIOVIRUS

Seven monoclonal neutralizing antibodies for type 2 Sabin vaccine strain poliovirus were prepared. It was possible to select mutants resistant to five of the antibodies at frequencies between one per 10^3 and 10^4 PFU of parental virus. The remaining two monoclonal antibodies were recloned and failed to yield mutants, even when selecting from $10^{10.7}$ PFU. This suggests that the antibodies may be directed against a site on the virus which is essential for infectivity.

Resistant mutants selected with type 2 specific monoclonal antibodies fell into a single group with five distinct subgroups on the basis of their resistance to the antibodies. A total of 150 plaques were tested. All had mutations within a region corresponding to site 1 of type 3 poliovirus, between amino acid residues 96 and 102 as shown in Fig. 2.

In addition to our own anti-type 2 monoclonal antibodies, 18 other samples were provided by other laboratories, giving a total of 25. The reactions of 14 of the 25 antibodies with the type 2 Sabin strain of poliovirus were affected by mutations within site 1, while the remaining 11 were not. Site 1 is therefore immunodominant in the induction of antibodies with neutralizing activity for type 2 poliovirus, but possibly less so than for type 3 poliovirus.

ANTIGENIC STRUCTURE OF TYPE 1 POLIOVIRUS

Seventeen monoclonal antibodies with neutralizing activity for the Sabin type 1 vaccine strain were prepared, and resistant mutants were selected. The frequencies of mutant isolation varied between one per $10^{3.5}$ and 10^6 PFU of parental virus, and an analysis of the patterns of reaction of over 400 mutant plaques led to the tentative conclusion that the antibodies were recognizing four independently mutable antigenic sites, which were designated A, B, C, and D. In all cases examined, the resistance of a mutant to neutralization by an antibody could be attributed to a failure to bind the antibody.

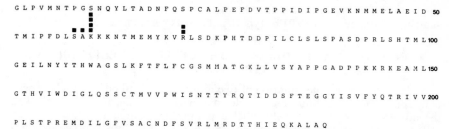

G L P V M N T P G S N Q Y L T A D N F Q S P C A L P E F D V T P P I D I P G E V K N M M E L A E I D 50

T M I P F D L S A K K K N T M E M Y K V R L S D K P H T D D P I L C L S L S P A S D P R L S H T M L 100

G E I L N Y Y T H W A G S L K F T F L F C G S M M A T G K L L V S Y A P P G A D P P K K R K E A M L 150

G T H V I W D I G L Q S S C T M V V P W I S N T T Y R Q T I D D S F T E G G Y I S V F Y Q T R I V V 200

P L S T P R E M D I L G F V S A C N D F S V R L M R D T T H I E Q K A L A Q

FIG. 3. Amino acid sequence of VP3 of the Sabin type 1 poliovirus vaccine strain showing the location of amino acid substitutions for mutants of group A, resistant to Sabin specific antibodies (■).

All members of group A, which comprised six single mutants and three multiple mutants, proved to have amino acid substitutions within VP3 at amino acid residue 58, 59, 60, or 71 from the N terminus, as shown in Fig. 3. No mutants of group A had amino acid substitutions within the region of VP1 corresponding to site 1. This is strikingly similar to reports by other workers who selected mutants from the Mahoney strain of type 1 poliovirus and identified mutations at residues 60 and 73 from the N terminus of VP3 (1).

Group C consisted of three distinct single mutants and five multiple mutants which had been deliberately isolated by selection with antibodies of group A or D, followed by selection with an antibody of group C. A ninth mutant was selected from the Mahoney strain of poliovirus type 1. Five of these independently selected mutants, including the mutant Mahoney virus, had the same amino acid substitution, a change from serine to leucine at residue 220 of VP1. A sixth mutant had a substitution at residue 222 in VP1, and no mutant of this group had any alterations in the region of VP1 corresponding to site 1. There is tentative evidence that two of the three mutants with unlocated mutations have substitutions within VP2. The substitutions within VP1 also parallel those reported for Mahoney virus (1).

Mutations within the isolates of groups B and D have not yet been identified but they are presumably located within the regions of the structural portion of the genome which we have not as yet sequenced (the N-terminal 30 amino acids of VP1, the C-terminal 60 amino acids of VP2, and all of VP3 except for the central 60 amino acids). No mutations have been found within the region of VP1 corresponding to site 1 or within the sites identified for groups A and C.

The difference between the results reported for poliovirus serotypes 1 and 3 is thus not simply an artifact of the techniques used to isolate monoclonal antibodies and mutants but reflects a difference of an unknown nature between the serotypes. Other workers (1) have interpreted their findings to mean that antibodies binding to a site corresponding to site 1 of type 1 Mahoney virus select for mutations in VP3 and that antibodies binding to a site in VP1 at amino acids 70 to 75 from the N terminus select for mutations in VP1 at positions 220 to 222. This model was based on the binding of monoclonal antibodies to peptides in ELISA tests, however, and we believe that it is more likely that the mutations detected occur chiefly at the site to which the antibodies bind. A single antibody has been described by others which is unambiguously directed against the region of the VP1 of type 1 poliovirus corresponding to site 1 (14).

MONOCLONAL ANTIBODIES TO THE CELLULAR RECEPTOR FOR POLIOVIRUS

We have described the isolation of two anticellular antibodies which appear to be specific for the cellular receptor for poliovirus (10). The antibodies blocked the adsorption of radioactively labeled poliovirus to monolayers of HEp-2c cells. They also protected cell monolayers from infection by all three serotypes of poliovirus while having no effect on the multiplication of any other viruses tested, including 11 enteroviruses and 10 other viruses. The extreme specificity of the antibodies suggests that they are directed against the receptor site for poliovirus (9). The concentration of antibody required to protect cells from infection was independent of the strain or serotype of poliovirus used and was the same for two cell lines of human origin and two derived from African green monkey, suggesting that the antigen is well conserved across these species.

Rabbits immunized with the anti-poliovirus receptor antibodies produced sera which inhibited the protective effect for HEp cells at dilutions of 1:100, consistent with the view that the sera contain anti-idiotype antibodies specific for the anti-receptor antibody binding site. Such anti-idiotypic antibodies might be expected to have a structure similar to that of the cellular receptor site and so react with native virions, but the sera failed to react with or neutralize poliovirus.

The anti-receptor antibodies were each able to reduce the replication of polioviruses of all three serotypes by a factor of 10^6, but virus which had grown in the presence of the antibodies proved to be of the parental type. Virus is currently being passed continually in the presence of the antibodies in an attempt to isolate virus receptor mutants. Finally, attempts to identify the receptor site by immune precipitation have so far failed, probably as a result of degradation during solubilization of the cell membranes.

The studies described here have thus identified several regions on poliovirions which are important in the neutralization of virus by antibodies. There are unexpected but real differences among the three serotypes. Monoclonal antibodies against the cellular receptor site for polioviruses have been isolated and will be of value in studies of the nature of the site.

LITERATURE CITED

1. **Diamond, D. C., B. A. Jameson, J. Brown, N. Kohara, S. Abe, H. Itoh, T. Komatsu, M. Arita, S. Kuge, A. D. M. E. Osterhaus, R. Crainic, A. Nomoto, and E. Wimmer.** 1985. Antigenic variation and resistance to neutralization in poliovirus type 1. Science **229**:1090–1093.

2. **Emini, E. A., Y. K. Shaw, A. J. Lewis, R. Craininc, and E. Wimmer.** 1983. Functional basis of poliovirus neutralization determined with monospecific neutralizing antibodies. J. Virol. **46**:466–474.

3. **Emini, E. A., B. A. Jameson, and E. Wimmer.** 1983. Priming for and induction of antipoliovirus neutralising antibodies by synthetic peptides. Nature (London) **304**:699–702.

4. **Evans, D. M. A., P. D. Minor, G. C. Schild, and J. W. Almond.** 1983. Critical role of an eight amino acid sequence of VP1 in neutralization of poliovirus type 3. Nature (London) **304**:454-462.

5. **Ferguson, M., P. D. Minor, D. I. Magrath, Y.-H. Qi, M. Spitz, and G. C. Schild.** 1984. Neutralization epitopes on poliovirus type 3 particles: an analysis using monoclonal antibodies. J. Gen. Virol. **65**:197–201.

6. **Ferguson, M., Y.-H. Qi, P. D. Minor, D. I. Magrath, M. Spitz, and G. C. Schild.** 1982. Monoclonal antibodies specific for the Sabin strain of poliovirus 3. Lancet **ii**:122–124.
7. **Hogle, J. M.** 1982. Preliminary studies of crystals of poliovirus type 1. J. Mol. Biol. **160**:663–666.
8. **Kitamura, N., B. L. Semler, P. G. Rothberg, G. R. Larsen, C. J. Adler, A. J. Dorner, E. A. Emini, R. Hanecak, J. L. Lee, S. Vander Werf, C. Anderson, and E. Wimmer.** 1981. Primary structure, gene organization and polypeptide expression of poliovirus RNA. Nature (London) **291**:547–553.
9. **Lonberg-Holm, K., R. L. Crowell, and L. Philipson.** 1976. Unrelated animal viruses share receptors. Nature (London) **259**:679–681.
9a.**Minor, P. D., D. M. A. Evans, M. Ferguson, G. C. Schild, G. D. Westrop, and J. W. Almond.** 1985. Principal and subsidiary antigenic sites of VP1 involved in the neutralization of poliovirus type 3. J. Gen. Virol. **65**:1159–1165.
10. **Minor, P. D., P. A. Pipkin, D. Hockley, G. C. Schild, and J. W. Almond.** 1984. Monoclonal antibodies which block cellular receptors of poliovirus. Virus Res. **1**:203–212.
11. **Minor, P. D., G. C. Schild, J. Bootman, D. M. A. Evans, M. Ferguson, P. Reeve, M. Spitz, G. Stanway, A. J. Cann, R. Hauptmann, L. D. Clarke, R. C. Mountford, and J. W. Almond.** 1983. Location and primary structure of a major antigenic site for poliovirus neutralization. Nature (London) **301**:674–679.
12. **Stanway, G., P. J. Hughes, R. C. Mountford, P. Reeve, P. D. Minor, G. C. Schild, and J. W. Almond.** 1984. Comparison of the complete nucleotide sequences of the genomes of the neurovirulent poliovirus P3/Leon/37 and its attenuated Sabin vaccine derivance P3/Leon 12a$_1$b. Proc. Natl. Acad. Sci. USA **81**:1539–1543.
13. **Toyoda, H., M. Kohand, Y. Kataoka, T. Suganuma, T. Omata, N. Imuta, and A. Nomoto.** 1984. Complete nucleotide sequence of all three poliovirus serotype genomes: implications for genetic relationship, gene function and antigenic determinants. J. Mol. Biol. **174**:561–570.
14. **Wychowski, C., S. Vander Werf, O. Siffert, R. Crainic, P. Burneau, and M. Girard.** 1983. A poliovirus type 1 neutralization epitope is located within amino acid residues 93 to 104 of viral capsid polypeptides VP1. EMBO J. **2**:2019–2023.

Anti-Idiotypic Antibodies against Anti-Coxsackievirus Type B4 Monoclonal Antibodies

PATRICK R. McCLINTOCK,†* BELLUR S. PRABHAKAR, AND ABNER LOUIS NOTKINS

Laboratory of Oral Medicine, National Institute of Dental Research, Bethesda, Maryland 20892

We have made anti-idiotypic antibodies in rabbits against three monoclonal neutralizing antibodies with specificities for independent epitopes on coxsackievirus type B4. Each of the anti-idiotypic antibodies was shown to be specific for the immunizing antibody and to inhibit the function of that antibody. These anti-idiotypic antibodies were then tested for their ability to recognize receptors for coxsackievirus type B4. None of the anti-idiotypes bound to the surface of coxsackievirus type B4 receptor-positive cells, nor did they inhibit the attachment of radiolabeled coxsackievirus type B4 to cells. The anti-idiotypic antibodies did induce an anti-anti-idiotypic antibody response in mice when measured by a competitive inhibition assay, but little if any increase in virus-neutralizing activity was found. Our results suggest that not all monoclonal neutralizing antibodies interact with cellular attachment sites on viruses and that not all anti-idiotypic antibodies are capable of eliciting a strong antiviral response.

In the relatively few years since the publication of the network hypothesis (7), there has been intensive study of the idiotypic repertoires of mice and humans (1). One avenue of investigation that has recently been opened is the use of internal image anti-idiotypic antibodies as probes for viral receptors and as potential immunogens for the production of antiviral antibodies (10, 13, 15, 18). The use of anti-idiotypic antibodies in this fashion (Fig. 1) is based upon the premise that the antigen-binding domain (paratope) of an antibody is a structure complementary to the antigen it recognizes (i.e., it is a three-dimensional mirror image of the antigen). An antibody (Ab1) can be used as an immunogen to elicit anti-idiotypic antibodies (Ab2), and some of these Ab2 molecules will possess paratopes complementary to the paratope of Ab1. Thus, some of the Ab2 antibodies will present paratopes which may resemble the antigenic determinant recognized by Ab1. If Ab1 is an antiviral antibody with specificity for a structure on the virus through which receptor attachment occurs, then some of the paratopes on Ab2 antibodies may resemble that structure to the extent that they can bind to the receptor. By extension, some anti-anti-idiotypic antibodies (Ab3) made by immunizing with Ab2 may be able

---

†Current address: American Type Culture Collection, Rockville, MD 20852.

36

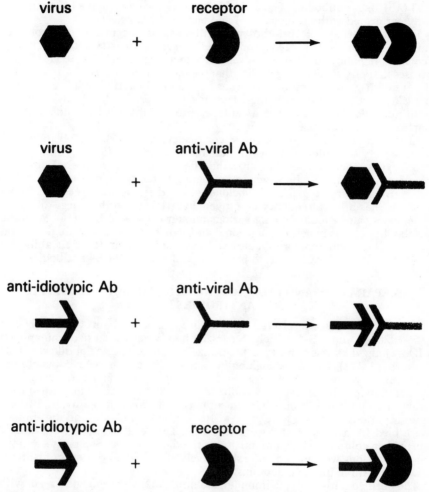

FIG. 1. Strategy for generation of anti-idiotypic antibodies with antiviral receptor specificity. Both the virus and the paratope of the anti-idiotypic antibody can bind to the paratope of the antiviral antibody if they present three-dimensional mirror images of the antiviral paratope. If the antiviral antibody is against a structure on the virus involved in receptor attachment, the anti-idiotypic antibody may present a mirror image of the binding site on the receptor and thus be able to react with the receptor.

to bind the original antigen recognized by Ab1, since they will possess paratopes complementary to the Ab2 paratopes that are themselves complementary to the Ab1 paratope. Thus, some anti-idiotypic antibodies may be able to serve as vaccines for some antigens by virtue of their ability to elicit antigen-binding

Ab3 (Ab1-like) antibodies. The above types of experiments have been successfully performed with hormones (16) and bacterial antigens (17).

Recently, the use of anti-idiotypic antibodies as vaccines or as probes of immune responses has been productive in several virus systems, including reovirus (4), poliovirus (18), and several enveloped viruses (3, 6, 8, 9, 11, 15). For viruses that possess defined proteins that mediate the virus-receptor interaction (i.e., reovirus), antireceptor activity has been found in anti-idiotypic antibodies made against antiviral antibodies. Thus far, however, similar results obtained with anti-idiotypic antibodies as anti-receptor antibodies have not been reported for picornaviruses.

Like certain other picornaviruses, coxsackievirus type B4 (CB4) induces a wide variety of clinical diseases in humans and animals (5, 19). Although no proof exists, one can speculate that the different disease manifestations may result from variants of CB4 which can recognize tissue-specific receptors on host cells. Using a large panel of monoclonal antibodies, we have recently shown that CB4 has a high frequency of antigenic variation and that large antigenic differences can be found among various human isolates of CB4 (14). With the availability of this panel of monoclonal antibodies, we have now made several polyclonal anti-idiotypic antibodies and have tested them for their ability to bind to receptors for CB4 and to prime for and elicit anti-CB4 antibodies.

PREPARATION AND CHARACTERIZATION OF ANTI-IDIOTYPIC ANTIBODIES

For these studies, we selected three mouse monoclonal anti-CB4 antibodies (356-1, 204-4, and 183-4) which recognized highly conserved determinants on CB4 as judged by their ability to neutralize a high percentage of different CB4 isolates (14, 14a). The immunoglobulins were purified from hybridoma culture medium by protein A-Sepharose chromatography, and Fab fragments were prepared by papain digestion. Rabbits (800-g male New Zealand whites) were immunized with 100 µg of Fab fragment in complete Freund adjuvant. Booster immunizations of 100 µg were given at 10- to 14-day intervals, and the animals were bled from the ear at 3-week intervals. The resulting sera were affinity purified on columns of the immunizing antibodies bound to Sepharose, and the bound fractions were eluted and exhaustively absorbed on columns of isotype-matched BALB/c immunoglobulins.

The purified anti-idiotypic immunoglobulins (Ab2) were shown to react with the appropriate immunizing antibody (Ab1), but not with the other anti-CB4 antibodies (Table 1). Plates were coated with each of the purified anti-idiotypic antibodies, 1 µg per well (i.e., anti-183-4 and anti-204-4), and incubated with a panel of monoclonal anti-CB4 antibodies representing different immunoglobulin classes and subclasses. The binding of radiolabeled 183-4 and 204-4 to their respective anti-idiotypic antibodies was then measured. Only the homologous combinations of antibody and its anti-idiotypic partner gave significant inhibition of binding. That is, unlabeled 183-4 blocked the binding of labeled 183-4 to anti-183-4-coated plates, but did not affect the binding of 204-4 to its anti-idiotypic antibody. Similarly, 204-4 blocked the binding of labeled 204-4 to anti-204-4-coated plates, but did not affect the binding of 183-4 to anti-183-4.

TABLE 1. Specific binding of anti-idiotypic antibodies (Ab2) to immunizing antibodies (Ab1) demonstrated by inhibition tests

Monoclonal antibody	Isotype[a]	% Inhibition[b] of binding of:	
		183-4	204-4
183-4	IgG1	71	17
204-4	IgG2a	5	91
2-2	—[c]	9	10
9-4	IgG3	10	18
38-1	IgG2a	6	16
59-3	—	5	6
86-3	IgG2b	6	14
102-3	—	7	16
128-2	IgM	10	18
187-1	IgG2a	11	0
204-3	IgG	9	0
208-2	IgG2b	9	9
215-2	IgG2a	9	5
356-1	IgG2a	9	8
Med	—	0	0

[a] Determined by immunodiffusion using specific typing sera.
[b] Immulon II Removawell strips were coated with anti-idiotypic antibodies. Monoclonal antibodies were added to the wells and incubated for 1 h at 37°C. The wells were washed and then incubated with ^{125}I-labeled 183-4 or 204-4 for 1 h. After extensive washing, the wells were counted and the percent inhibition relative to medium alone was calculated.
[c] Isotype was not determined.

None of the other monoclonal antibodies inhibited the binding of 183-4 or 204-4 to their respective anti-idiotypic antibodies.

The above results were confirmed in assays measuring virus neutralization (14) and immunoprecipitation of radiolabeled virus (data not shown). Several monoclonal anti-CB4 antibodies were incubated with the anti-idiotypic antibodies and then tested for their ability to neutralize CB4 or to precipitate labeled CB4. In both types of assays, the anti-idiotypic antibodies selectively blocked the activity of the homologous monoclonal antibody. These results, together with the data in Table 1, indicated that our anti-idiotypic antibody preparations contained antibodies against the paratopes of the respective monoclonal antibodies. This raised the possibility that the anti-idiotypic antibodies might also possess antireceptor activity.

RECEPTOR AND CELL-BINDING ACTIVITY OF ANTI-IDIOTYPIC ANTIBODIES

To test for anti-receptor antibodies, we used Buffalo green monkey kidney (BGMK) cells in single-cell suspensions and measured the ability of anti-idiotypic antibodies to inhibit the binding of radiolabeled CB4 virions. The virus-attachment assays were done essentially as described previously (12) except that they were performed at 20°C with a virus particle to cell ratio of approximately 1,000:1. Normal rabbit immunoglobulin G (IgG) (200 μg/ml),

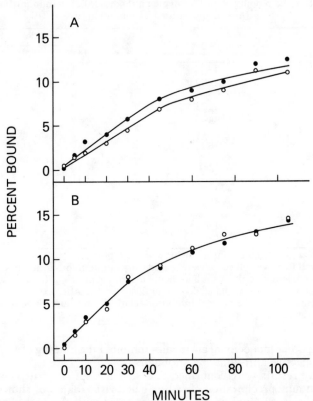

FIG. 2. Failure of anti-idiotypic antibodies to inhibit the binding of CB4 to cells. BGMK cells were suspended in Eagle minimum essential medium with 5% calf serum at a concentration of approximately 5×10^6 cells per ml. Anti-204-4 (A) or anti-356-1 (B) was added to the suspensions at 150 µg/ml and 130 µg/ml, respectively. Normal rabbit IgG (200 µg/ml) was used as a control. The suspensions were stirred for 1 h at 37°C and then cooled to 20°C, and radiolabeled CB4 was added. At the indicated times the suspensions were sampled and the percentage of the virus bound was determined. Symbols: ●, anti-idiotypic antibody; ○, normal rabbit IgG.

anti-204-4 (150 µg/ml), or anti-356-1 (130 µg/ml) was added to the cell suspension 1 h before the addition of the virus. After addition of the virus, samples of the suspension were taken at appropriate times, the bound and free virus was measured, and the percent bound was calculated. Figure 2 shows that neither anti-204-4 (A) nor anti-356-1 (B) significantly affected the binding of CB4 to BGMK cells. Neither the rate of attachment nor the final extent of attachment differed between the cells treated with anti-idiotypic antibodies and those treated with normal IgG. Thus, despite the fact that these anti-idiotypic antibodies are

TABLE 2. Expression of idiotypes in mice immunized with anti-idiotypic antibodies

Immunization[a]	% Inhibition[b] of binding of:	
	183-4	204-4
None	25[c]	25
Anti-183	57	28
Anti-204	35	45
Anti-356	30	30

[a] BALB/c female mice were immunized intraperitoneally with anti-idiotypic antibodies (10 μg per mouse) in saline on days 0, 8, and 20 and were bled on day 27.
[b] Percent inhibition of binding of ^{125}I-labeled monoclonal antibody 183-4 or 204-4 to plates coated with their respective anti-idiotypic antibodies by sera from unimmunized animals and from animals immunized with the anti-idiotypic antibodies.
[c] Numbers indicate the mean value for duplicate samples from five mice per group.

apparently directed against paratopes recognizing strongly conserved epitopes on CB4 (14a), no anti-CB4 receptor activity could be detected.

It was still possible that our anti-idiotypic antibodies might be capable of binding to receptors for CB4, but were unable to inhibit virus attachment. We therefore tested the ability of the anti-idiotypic antibodies to bind to the surface of BGMK cells (data not shown). In these assays, cells were grown on glass cover slips, and the cover slips were incubated with anti-idiotypic antibodies or normal rabbit IgG. The cover slips were then given a second incubation with ^{125}I-labeled goat anti-rabbit IgG, washed, and counted in a gamma counter. None of the anti-idiotypic antibodies (i.e., anti-183-4, anti-204-4, or anti-356-1) bound to BGMK cells above control levels. We conclude that these anti-idiotypic antibodies do not have antireceptor activity.

ANTI-ANTI-IDIOTYPIC ANTIBODIES

The failure of the anti-idiotypic antibodies to bind to receptors suggested that the monoclonal antibodies to which they were made did not recognize a receptor attachment site on the virus. Clearly, however, the monoclonal antibodies did neutralize CB4. If the anti-idiotypic antibodies (Ab2) contained the internal images of their respective anti-CB4 antibodies (Ab1), then anti-anti-idiotypic antibodies (Ab3) made by immunizing mice with Ab2 might possess Ab1 activity.

We immunized mice with anti-183-4, anti-204-4, and anti-356-1 and tested the sera of these mice for the presence of Ab3. The sera showed an increase in Ab3 activity against the specific immunizing Ab2, but little if any increase against other anti-idiotypic antibodies (Table 2). Sera from these mice when tested for anti-CB4 antibody also showed little if any increase in the neutralization titer relative to control mice (data not shown).

CONCLUSIONS

We have made anti-idiotypic antibodies against mouse monoclonal anti-CB4 antibodies and have tested them for their ability to recognize CB4 receptors.

These anti-idiotypic antibody preparations contained antibodies directed against the paratopes of their respective monoclonal antibodies; however, no evidence of antireceptor reactivity was found. Moreover, when the anti-idiotypic antibodies were used as immunogens in mice, the resulting sera contained anti-anti-idiotypic antibodies (Ab3) as detected by a competitive inhibition assay (Table 2). However, sera from these mice contained little if any anti-CB4 neutralizing activity. Furthermore, little or no increase in idiotype expression or neutralizing antibody was found in mice primed with Ab2 and then challenged with CB4 (data not shown).

Others (13) have shown that anti-idiotypic antibodies directed against antibodies to the sigma 1 hemagglutinin of reovirus type 3 can have potent antireceptor activity. Unlike reovirus, it is not known whether CB4, or any other picornavirus, possesses a specific structure (e.g., analogous to sigma 1) which mediates the attachment to receptors. In all probability, epitopes involved in picornavirus receptor binding constitute a small subset of neutralizing epitopes. It is even possible that receptor recognition by picornaviruses involves epitopes that are not involved in the neutralizing antibody response.

In our hands, immunization with anti-idiotypic antibodies did not induce or prime a strong neutralizing antibody response against CB4. Although some groups working with different viruses have reported successful manipulation of immune responses using anti-idiotypic antibodies (3, 9, 15, 18), neutralizing antibody responses have not always resulted (8). Furthermore, not all anti-anti-idiotypic antibodies would be expected to have antigen-binding (Ab1-like) capability (2), since not all will be of the "internal image" type and not even all internal image Ab3s will bind antigen (2). Perhaps for CB4, more monoclonal antibodies and their anti-idiotypes must be studied. Given the relative simplicity of the coxsackievirus capsid, one would expect that anti-idiotypic antibodies would provide a fruitful approach to receptor studies and immune modulation. These efforts may be aided by a detailed structural analysis of the exposed antigenic domains on CB4 and mapping of neutralizing and receptor attachment sites.

LITERATURE CITED

1. **Bona, C. A., and H. Kohler (ed.).** 1983. Immune networks. Ann. N.Y. Acad. Sci. vol. 418.
2. **Erlanger, B. F.** 1985. Anti-idiotypic antibodies: What do they recognize? Immunol. Today **6**:10–11.
3. **Ertl, H. C. J., and R. W. Finberg.** 1984. Sendai virus-specific T-cell clones: induction of cytolytic T cells by an anti-idiotypic antibody directed against a helper T-cell clone. Proc. Natl. Acad. Sci. USA **81**:2850–2854.
4. **Ertl, H. C. J., M. I. Green, J. H. Noseworthy, B. N. Fields, J. T. Nepom, D. R. Spriggs, and R. W. Finberg.** 1982. Identification of idiotypic receptors on reovirus-specific cytolytic T cells. Proc. Natl. Acad. Sci. USA **79**:7479–7483.
5. **Gear, J. H. S., and V. Measroch.** 1973. Coxsackievirus infections of the newborn. Prog. Med. Virol. **15**:42–62.
6. **Gheuns, J., D. E. McFarlin, K. W. Rammohan, and W. J. Bellini.** 1981. Idiotypes and biological activity of murine monoclonal antibodies against the hemagglutinin of measles virus. Infect. Immun. **34**:200–207.
7. **Jerne, N. K.** 1974. Towards a network theory of the immune responses. Ann. Immunol. (Paris) **125C**:373–389.
8. **Kennedy, R. C., K. Adler-Storthz, J. W. Burns, R. D. Henkel, and G. R. Dreesman.** 1984. Anti-idiotype modulation of herpes simplex virus infection leading to increased pathogenicity. J. Virol. **50**:951–953.
9. **Kennedy, R. C., K. Adler-Storthz, R. D. Henkel, Y. Sanchez, J. L. Melnick, and G. R. Dreesman.**

1983. Immune response to hepatitis B surface antigen: enhancement by prior injection of antibodies to the idiotype. Science **221**:853–855.

10. **Kennedy, R. C., and G. R. Dreesman.** 1985. Immunoglobulin idiotypes: analysis of viral antigen-antibody systems. Prog. Med. Virol. **31**:168–182.

11. **Liu, Y.-N., C. A. Bona, and J. L. Schulman.** 1981. Idiotype of clonal responses to influenza virus hemagglutinin. J. Exp. Med. **154**:1525–1538.

12. **McClintock, P. R., L. C. Billups, and A. L. Notkins.** 1980. Receptors for encephalomyocarditis virus on murine and human cells. Virology **106**:261–272.

13. **Nepom, J. T., H. L. Weiner, M. A. Dichter, M. Tardieu, D. R. Spriggs, C. F. Gramm, M. L. Powers, B. N. Fields, and M. I. Green.** 1982. Identification of a hemagglutinin-specific idiotype associated with reovirus recognition shared by lymphoid and neural cells. J. Exp. Med. **155**:155–167.

14. **Prabhakar, B. S., M. V. Haspel, P. R. McClintock, and A. L. Notkins.** 1982. High frequency of antigenic variants among naturally occurring human Coxsackie B_4 virus isolates identified by monoclonal antibodies. Nature (London) **300**:374–376.

14a. **Prabhakar, B. S., M. A. Menegus, and A. L. Notkins.** 1985. Detection of conserved and nonconserved epitopes on coxsackievirus B4: frequency of antigenic change. Virology **146**:302–306.

15. **Reagan, K. J., W. H. Wunner, T. J. Wiktor, and H. Koprowski.** 1983. Anti-idiotypic antibodies induce neutralizing antibodies to rabies virus glycoprotein. J. Virol. **48**:660–666.

16. **Sege, K., and P. A. Peterson.** 1978. Use of anti-idiotypic antibodies as cell-surface receptor probes. Proc. Natl. Acad. Sci. USA **75**:2443–2447.

17. **Urbain, J., M. Wikler, J. D. Franssen, and C. Collignon.** 1977. Idiotypic regulation of the immune system by the induction of antibodies against anti-idiotypic antibodies. Proc. Natl. Acad. Sci. USA **74**:5126–5130.

18. **Utydehaag, F. G. C. M., and A. D. M. E. Osterhaus.** 1985. Induction of neutralizing antibody in mice against poliovirus type II with monoclonal anti-idiotypic antibody. J. Immunol. **134**:1225–1229.

19. **Yoon, J. W., M. Austin, T. Onodera, and A. L. Notkins.** 1979. Virus-induced diabetes mellitus: isolation of a virus from the pancreas of a child with diabetic ketoacidosis. N. Engl. J. Med. **300**:1173–1179.

Biochemical Characterization of Polyomavirus-Receptor Interactions

R. A. CONSIGLI, G. R. GRIFFITH, S. J. MARRIOTT, AND
J. W. LUDLOW

*Section of Virology and Oncology, Division of Biology, Kansas State University,
Manhattan, Kansas 66506*

Polyoma virions have different attachment proteins which are
responsible for hemagglutination of erythrocytes and attachment to
cultured baby mouse kidney cells (MKC). Virion binding studies
demonstrated that MKC possess specific (productive infection) and
nonspecific (nonproductive) receptors. Empty polyoma capsids have
hemagglutination activity and bind to nonspecific MKC receptors,
but are not capable of competing for specific virion cell receptors or
preventing productive infection. Isoelectric focusing of the virion
major capsid protein, VP1, separated this protein into six species (A
through F) and the empty capsid VP1 into four species. These
species had identical amino acid sequences, but differed in degree of
phosphorylation, acetylation, or both. Evidence based upon precipitation with specific antisera supports the view that VP1 species E is
required for specific absorption and that D and F are required for
hemagglutination. Monopinocytotic vesicles containing ^{125}I-labeled
polyoma virions were isolated from infected MKC. The noncleavable heterobifunctional photoreactive cross-linker N-hydroxysuccinimidyl 4-azidobenzoate was used to bind the cell receptors
covalently to VP1 attachment protein. This complex was solubilized
from the monopinocytotic vesicles and identified in Western blots of
sodium dodecyl sulfate-polyacrylamide gels using anti-idiotype antibodies prepared against a monoclonal antibody which specifically
blocked virus attachment.

BACKGROUND

The late region of polyomavirus DNA consists of 2,366 base pairs which code
for the virus structural proteins VP1, VP2, and VP3 (12). The major capsid
protein VP1 is encoded at the 3' end of the late region and is translated from a
16S mRNA, whereas the minor capsid proteins VP2 and VP3 are encoded at the
5' end and are translated from 19S and 18S mRNAs (14). These three proteins
make up the structural units of the virus, and they are also essential for
attachment to sites on host cells, for agglutination of erythrocytes, and in DNA
packaging and virus assembly. In contrast to the detailed knowledge of the late
stages of viral infection, little is known about virion adsorption and entry of
polyomavirus into host cells.

Early studies with polyomavirus demonstrated that the hemagglutinin was
part of the capsid structure (11, 13). Other studies also suggested that the virion

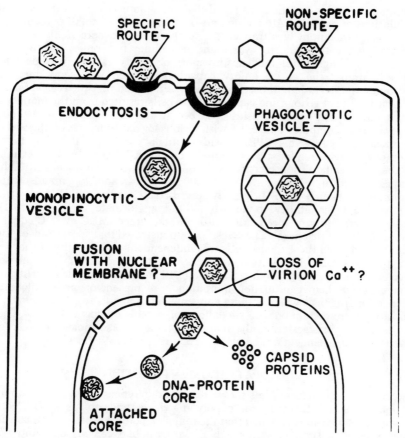

FIG. 1. Model of polyoma virion attachment, penetration, and nuclear entry (5, 8, 15, 18, 24).

hemagglutinin and attachment protein for host cells were different entities, since it was reported by several laboratories (10, 16, 21) that nonspecific cellular and serum inhibitors could prevent hemagglutination but not infection. Recent electron microscopic (18) and biochemical studies (4, 5, 15) have revealed that polyoma virions and empty capsids have different attachment proteins and adsorb to different cell receptors. Figure 1 illustrates the early events of the infection of primary mouse kidney cells (MKC) by polyomavirus as characterized by electron microscopy (18), binding (5), and nuclear uncoating studies (18, 24). In both permissive and nonpermissive cells, the virus penetrates by endocytosis into the cytoplasm in monopinocytotic vesicles. These vesicles then migrate to the nucleus, where the virus is uncoated and the replication process can begin. Mackay and Consigli (18) also compared the fate of polyoma virions and empty capsids and demonstrated distinct differences between these two populations of particles. After virion adsorption to cells, a small percentage of the original inoculum was transported to the nucleus where replication was initiated, while the majority was found in cellular lysosomes, where it was degraded. No significant nuclear transport of empty capsids was detected. Instead, almost the entire empty capsid population was sequestered by lysosomes and degraded. Subsequently, it was reported that the reason for this differential fate of virions and capsids within cells resulted from the manner by which virions and empty capsids adsorbed to the cell surface (5). Although empty capsids competed efficiently with virions for receptor sites on the surface of guinea pig erythrocytes (GPE), capsids could not compete with virions for the receptor sites required for successful infection of MKC. Recently, the maximum number of specific virion receptor sites on the surface of quiescent MKC has been estimated to be approximately 10,000 per cell. In addition, an adsorption mutant of polyoma (Py 235), originally described by Basilico and DiMayorca (3), which lacked the ability to adsorb to and agglutinate GPE nevertheless possessed an ability to specifically adsorb to and infect MKC at 32°C equal to that of wild-type virions (5). These results demonstrated that specific adsorption to and infection of these mouse cells was independent of the ability of the virus to agglutinate GPE. In addition, we were able to demonstrate with antisera directed against cleavage products of virion VP1 that the antigenic determinants responsible for hemagglutination inhibition were present on an 18-kilodalton polypeptide fragment whereas those antigenic determinants responsible for neutralization were present on a 16-kilodalton polypeptide fragment (1, 6, 20). These results strongly suggest that the majority of the diverse biological functions observed were dependent on the major structural protein VP1.

VIRION ATTACHMENT PROTEIN

Recently, our laboratory has shown distinct differences between the subpopulations of the structural proteins of polyoma virions and capsids (4). Isoelectric focusing gels of virions revealed that the major structural protein, VP1, consists of six species, designated A through F, with pIs between pH 6.75 and 5.75. Capsids were found to have only four VP1 species with pIs between pH 6.60 and 5.75 and to lack two species (Fig. 2). The differences in pIs of the VP1 species conceivably resulted from different degrees and types of posttranslational modifications of the VP1 species. Three of the virion VP1 species were differentially phosphorylated (D, E, and F), and two of the species were acetylated (C and D). Two of the capsid VP1 species were phosphorylated (D and F), and D, as well as the other two species, was also acetylated. Our laboratory is currently investigating other VP1 modifications, namely, methylation and sulfation. Peptide mapping of the virion VP1 species has indicated no differences in amino acid sequence (2). The variety of presumably host-contributed posttranslational modifications of VP1 suggests that such modifications allow a number of different functions to be expressed and permit the virus to use its limited genetic coding capacity efficiently. Multiple species of polyoma virion VP1 differing in their pIs have been reported previously (17, 23). It is of considerable interest that O'Farrell and Goodman (22) also reported that simian virus 40 virions

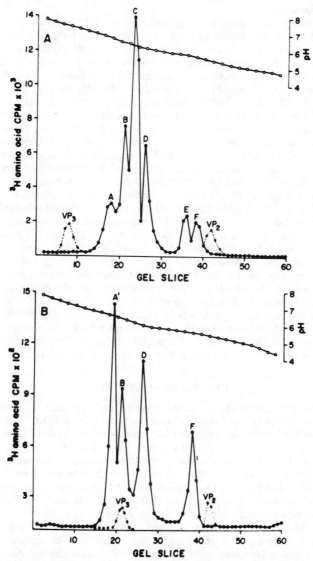

FIG. 2. Isoelectric focusing of ³H-amino acid-labeled virions and capsids. Polyoma virions and capsids were purified from the lysates of infected cells grown in the presence of ³H-amino acids during the period of virus infection. ³H-amino acid-labeled virions (A) and capsids (B) were then analyzed by isoelectric focusing and SDS-PAGE (4).

possess six distinct VP1 species. This pattern of six VP1 species may be common to all papovaviruses.

Our experiments have allowed us to tentatively assign functions to the various species of VP1. Polyoma virion VP1 may serve at least five distinct roles, namely, structural functions (4, 8, 9, 25), regulation of early transcription (4, 7), protease activity (1, 6, 20), hemagglutination, and host-cell attachment (4, 5). This article deals with the last two functions. Experimental data indicate that three of the virion VP1 species (D, E, and F) are involved in these activities. VP1 species D and F appear to be common to both virions and empty capsids, although the capsid species may differ slightly from their virion counterparts. It is also possible that not all modifications observed dictate functional changes; however, these modifications may allow conformational diversity of the VP1 molecules required for assembly of virus particles. The additional biological functions attributed to some of these species could be dependent on the presence of unidentified modifications or in many instances could be independent of such alterations. The functional identity of virion species D and F was indicated by immune precipitation experiments. Antibodies which specifically inhibited virion or capsid adsorption to and agglutination of GPE recognized only VP1 species D and F. Thus it appears that D and F are involved with hemagglutination. Virion VP1 species E appears to be required for successful virus adsorption to specific cellular receptors, which results in infection of MKC. Thus, VP1 species E is the polyoma virion attachment protein. Two lines of evidence support this contention. First, VP1 species E is found only on virions, never on empty capsids. Since, as discussed above, empty capsids do not compete with virions for specific cellular receptors (5), it seems logical that capsids would lack the corresponding protein present on virions. Second, antibodies whose only known activity was to inhibit virion adsorption to MKC, but not to GPE, specifically recognized distinct antigenic determinants only on VP1 species E in immune precipitation experiments (4).

CELLULAR RECEPTOR

Our laboratory also initiated studies to identify the cellular receptor(s) which is involved in initiating polyoma productive infection of MKC. We elected to approach this problem by isolation of monopinocytotic vesicles containing polyomavirus from the cytoplasm of infected MKC. Such virus-containing vesicles are derived from the plasma membrane and should provide an enriched source of specific cellular receptor(s) united with polyomavirus attachment protein(s). MKC were infected with purified ^{125}I-labeled polyomavirus at 4°C to allow maximum adsorption, and then cultures were shifted to 37°C for 1 h to allow virus penetration. Cells were disrupted, and monopinocytotic vesicles containing ^{125}I-labeled virus were isolated as described by Griffith and Consigli (15). Figure 3 illustrates the sucrose gradient isolation of two peaks of vesicles containing virions. In addition, the plasma membrane marker alkaline phosphatase was principally associated with peak I of the gradient. The lysosomal marker β-glucuronidase, however, showed most of its activity on the shoulder of peak II, toward the top of the gradient, with little activity in peak I. Additional experiments (data not shown) showed that the plasma membrane marker 5-nucleotidase was associated with both peaks I and II. These results suggested

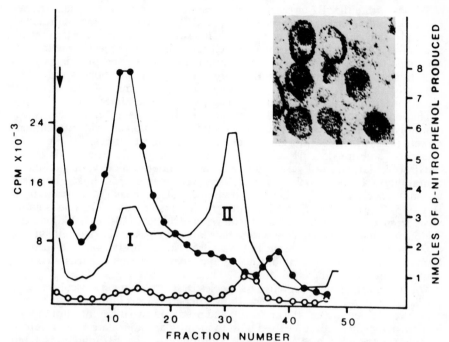

FIG. 3. Sedimentation pattern of monopinocytotic vesicles containing polyoma virions and their associated enzyme activities. The cytoplasmic fraction containing 6.94×10^5 cpm of ^{125}I-labeled polyoma virions was analyzed on a 15 to 45% sucrose gradient (the top of the gradient is to the right). Samples of 0.1 ml were taken from each fraction and assayed for alkaline phosphatase and β-glucuronidase. Activities were plotted as nanomoles of product formed during the incubation. Symbols: Solid line, ^{125}I-labeled virus; ●, alkaline phosphatase; ○, β-glucuronidase. The arrow indicates the sedimentation of purified ^{125}I-labeled polyoma virus. Insert is a thin-section electron micrograph showing vesicles from peak I containing virions (15).

that the virus recovered in both peaks may be associated with plasma membrane, rather than the lysosomes. Thin-section electron microscopy of the isolated vesicles revealed intact virions having a diameter of 40 to 42 nm, similar to purified virions. The monopinocytotic vesicles containing virions had an average diameter of 50 to 60 nm (Fig. 3 insert). It is worthy of mention that some virions appeared to remain attached to the inside of the vesicle membrane (previously the outside of the plasma membrane), and this may reflect the state of attachment of the virion to its cellular receptor.

A panel of monoclonal antibodies to polyoma VP1 have been characterized in our laboratory (19). Two of these monoclonal antibodies (C10 and D3) are specific for VP1 species B and C (nonattachment species), and two other monoclonal antibodies (E7 and G9) are specific for the VP1 attachment species D, E, and F. Treatment of virions with either E7 or G9 monoclonal antibodies to the attachment species reduced virion internalization by 99%, while C10 and D3 monoclonal antibodies reduced virion internalization by only 8%. These observations indicate that the E7 and G9 monoclonal antibodies neutralized the virion attachment proteins. The combined treatment of MKC with neuraminidase (0.15 U) and infecting virions with mixed gangliosides (2 mg/ml) reduced virion internalization by only 90%. This may suggest that the MKC receptor(s) involves neuraminidase-sensitive and -resistant residues, both of which are needed for maximum infection.

The noncleavable heterobifunctional photoreactive cross-linker N-hydroxysuccinimidyl 4-azidobenzoate was used to visualize the union of virion attachment protein (VP1) to cell receptor in the isolated monopinocytotic vesicles. Figure 4A illustrates sodium dodecyl sulfate polyacrylamide gel electrophoresis (SDS-PAGE) of isolated peak I vesicles (lane 2) and isolated peak II vesicles (lane 3) containing ^{125}I-labeled polyoma virions, as well as free, cross-linked polyoma virions recovered from Nonidet P-40–treated vesicles in peak I (lane 4) and peak II (lane 5). The SDS-PAGE and autoradiography of isolated vesicles and isolated virus from these vesicles revealed a new band in the 110- to 120-kilodalton region. The band was detected with either labeled virus or labeled cell surface proteins, but not in the absence of N-hydroxysuccinimidyl 4-azidobenzoate. Pretreatment which prevented binding and internalization of virus also prevented formation of the cross-linked 120-kilodalton band.

The use of attachment protein VP1 species D, E, and F in affinity columns for the isolation of MKC receptor(s) was impractical because of limiting quantities. Thus, we decided to alter the approach and prepared two different immunological reagents which could be used both to identify and to isolate MKC receptor protein(s). The first was a polyclonal antibody to the SDS-PAGE–isolated cross-linked (virion-MKC receptor) 120-kilodalton band; the second was a polyclonal anti-idiotype antibody to Fab fragments of E7 monoclonal antibody (anti-VP1 attachment D, E, and F species). To separate anti-idiotype Fab antibodies from other anti-mouse immunoglobulin G antibodies we used a staphylococcal protein A column bound with normal BALB/c mouse immunoglobulin G. The anti-idiotype was found to react positively with monoclonal antibody E7, but not with polyoma proteins, as determined by enzyme-linked immunosorbent assay or by immunoprecipitation. In addition, the anti-idiotype was capable of attaching to the surface of MKC, to compete with polyomavirus for MKC receptors and to prevent infection determined by indirect immunofluorescence. These immunological reagents were also tested on Western blots to determine their specificity (Fig. 4, panel B). The antibody to the isolated 120-kilodalton band was found to react specifically with polyoma VP1 protein(s), namely, dimer, VP1, VP1 fragments (18 and 16 kilodaltons), and the cross-linked 120-kilodalton band, but not with VP2 and VP3 structural proteins (lane 2). The anti-idiotype antibody did not react with the SDS-PAGE–separated polyoma proteins (lane 3). Both immunological reagents showed reactivity to seven uninfected MKC membrane proteins with molecular weights of 12,000, 23,000, 28,000, 30,000, 38,000, 52,000, and 110,000 (lanes 5 and 6), with the strongest reacting band at 28,000. These multiple bands might reflect the dissociation of the MKC polyoma receptor under the denaturing and reducing conditions of the SDS-PAGE. Currently we are optimistic that these two immunological reagents will allow

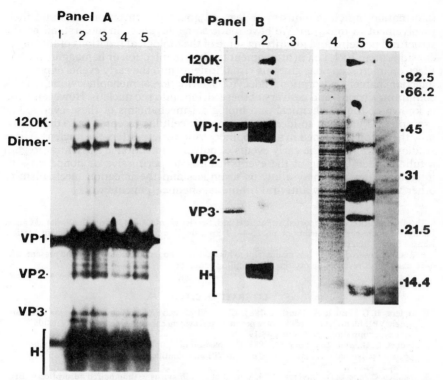

FIG. 4. SDS-PAGE analysis of reduced proteins from virions and MKC membranes. (A) Lane 1, ^{125}I-labeled virions, peak I (see Fig. 3); lane 2, isolated monopinocytotic vesicles containing cross-linked ^{125}I-labled virions, peak I (see Fig. 3); lane 3, isolated monopino-cytotic vesicles containing cross-linked ^{125}I-labeled virions, peak II (see Fig. 3); lane 4, cross-linked ^{125}I-labeled virions recovered from peak I of vesicles; lane 5, cross-linked ^{125}I-labeled virions recovered from peak II of vesicles. (B) Lane 1, polyoma virion proteins stained with Coomassie blue; lane 2, Western blot of cross-linked virions recovered from vesicles and probed with ^{125}I-labeled anti-120-kilodalton immunoglobulin G; lane 3, Western blot of virions probed with ^{125}I-labeled anti-E7 idiotype Fab; lane 4, MKC membrane protein stained with Coomassie blue; lane 5, Western blot of MKC membranes probed with ^{125}I-labeled anti-120-kilodalton immunoglobulin G; lane 6, Western blot of MKC membrane probed with anti-E7 idiotype Fab.

us to identify and to isolate sufficient quantities of the MKC polyoma receptor for biochemical characterization.

FUTURE WORK

The genome of polyomavirus has been sequenced and dissected and is almost completely mapped with genetic and biological markers. Despite a wealth of information, much remains to be learned about the virion proteins and their involvement in infection. We have made some advances in understanding the structural proteins and identifying some of the biological functions of the major capsid protein VP1, i.e., in attachment leading to infection or hemagglutination. Electron microscopy has given us visual insights into the early events of polyoma infection, namely, adsorption, endocytosis, formation of monopinocytotic vesicles containing virions, and delivery of these virions into the nucleus. However, little is known of the biochemical and biological mechanisms of these events. For example, we have yet to identify the specific cellular receptor or to determine how the virion contained in a monopinocytotic vesicle enters the nucleus. These studies dealing with the early events of polyoma infection are essential to gain a fuller understanding of the events leading to permissive or nonpermissive infection. They also may allow us to understand the infection mechanism of other viruses and how antiviral (immune, chemical) agents work.

This investigation was supported by Public Health Service grant CA-07139 from the National Cancer Institute. This paper is contribution no. 85-441-J from the Kansas Agricultural Experiment Station, Kansas State University.

R. A. C. wishes to dedicate this manuscript to his students and technical staff, past and present, who have made virus research an exciting and fulfilling endeavor.

LITERATURE CITED

1. **Anders, D. G., and R. A. Consigli.** 1983. Chemical cleavage of polyomavirus major structural protein VP1: identification of cleavage products and evidence that the recptor moiety resides in the carboxy-terminal region. J. Virol. **48:**197–205.
2. **Anders, D. G., and R. A. Consigli.** 1983. Comparison of nonphosphorylated and phosphorylated species of polyomavirus major capsid protein VP1 and identification of the major phosphorylation region. J. Virol. **48:**206–217.
3. **Basilico, C., and G. Di Mayorca.** 1974. Mutant of polyomavirus with impaired adsorption to BHK cells. J. Virol. **13:**931–934.
4. **Bolen, J. B., D. G. Anders, J. Trempy, and R. A. Consigli.** 1981. Differences in the subpopulation of the structural proteins of polyoma virions and capsids: biological functions of the VP1 species. J. Virol. **37:**80–91.
5. **Bolen, J. B., and R. A. Consigli.** 1979. Differential adsorption of polyoma virions and capsids to mouse kidney cells and guinea pig erythrocytes. J. Virol. **32:**679–683.
6. **Bolen, J. B., and R. A. Consigli.** 1980. Separation of neutralizing and hemagglutination-inhibiting antibody activities and specificity of antisera to sodium dodecyl sulfate-derived polypeptides of polyoma virions. J. Virol. **34:**119–129.
7. **Brady, J. N., C. Lavialle, and N. P. Salzman.** 1980. Efficient transcription of a compact nucleoprotein complex isolated from purified simian virus 40 virions. J. Virol. **35:**371–381.
8. **Brady, J. N., V. D. Winston, and R. A. Consigli.** 1977. Dissociation of polyomavirus by the chelation of calcium ions found associated with purified virions. J. Virol. **23:**717–724.
9. **Brady, J. N., V. D. Winston, and L. A. Consigli.** 1978. Characterization of a DNA-protein complex and capsomere subunits derived from polyomavirus by treatment with ethyleneglycol-bis-N-N'-tetraacetic acid and dithiothreitol. J. Virol. **27:**193–204.
10. **Cramer, R., and S. E. Stewart.** 1960. Zone electrophoresis studies on hemagglutination of a hemagglutinating and a masked strain of polyomavirus. Proc. Soc. Exp. Biol. **103:**697–700.
11. **Crawford, L. V., E. M. Crawford, and D. H. Watson.** 1962. The physical characteristics of polyomavirus. I. Two types of particles. Virology **18:**170–176.
12. **Deininger, P., A. Esty, P. La Porte, and T. Friedmann.** 1979. Nucleotide sequence and genetic organization of polyoma late region: features common to the polyoma early region and SV40. Cell **18:**771–779.
13. **Eddy, B. E., W. P. Rowe, J. W. Harley, S. E. Stewart, and R. J. Huebner.** 1958. Hemagglutination with the SE polyomavirus. Virology **6:**290–291.

14. **Fried, M., and B. Griffin.** 1977. Organization of the genomes of polyomavirus and SV40. Adv. Cancer Res. **24:**67–114.

15. **Griffith, G. R., and R. A. Consigli.** 1984. Isolation and characterization of monopinocytotic vesicles containing polyomavirus from the cytoplasm of infected mouse kidney cells. J. Virol. **50:**77–85.

16. **Hartley, J. W., W. P. Rowe, R. M. Chanock, and B. E. Andrews.** 1959. Studies of mouse polyomavirus infection. IV. Evidence for mucoprotein erythrocyte receptor in polyomavirus hemagglutination. J. Exp. Med. **110:**81–91.

17. **Hewick, R. M., M. D. Waterfield, L. K. Miller, and M. Fried.** 1977. Correlation between genetic loci and structural differences in the capsid proteins of polyomavirus plaque morphology mutants. Cell **11:**331–338.

18. **Mackay, R. L., and R. A. Consigli.** 1976. Early events in polyomavirus infection attachment, penetration, and nuclear entry. J. Virol. **19:**620–636.

19. **Marriott, S. J., and R. A. Consigli.** 1985. Production and characterization of monoclonal antibodies to polyomavirus major capsid protein VP1. J. Virol. **56:**365–372.

20. **McMillen, J., and R. A. Consigli.** 1977. Immunological reactivity of antisera to sodium dodecyl sulfate-derived polypeptides of polyoma virion. J. Virol. **21:**1113–1120.

21. **Mori, R., J. H. Schieble, and W. W. Ackermann.** 1962. Reaction of polyoma and influenza viruses with receptors of erythrocytes and host cells. Proc. Soc. Exp. Biol. **109:**685–690.

22. **O'Farrell, P. Z., and H. M. Goodman.** 1976. Resolution of simian virus 40 proteins in whole cell extracts by two dimensional electrophoresis: heterogeneity of the major capsid protein. Cell **9:**289–298.

23. **Ponder, B. A. J., A. K. Robbins, and L. V. Crawford.** 1977. Phosphorylation of polyoma and SV40 virus proteins. J. Gen. Virol. **37:**75–83.

24. **Winston, V. D., J. B. Bolen, and R. A. Consigli.** 1980. Isolation and characterization of polyoma uncoating intermediates from nuclei of infected mouse cells. J. Virol. **33:**1173–1181.

25. **Yuen, L. K. C., and R. A. Consigli.** 1985. Identification and protein analysis of polyomavirus assembly intermediates from infected primary mouse embryo cells. Virology **144:**127–138.

Viral Membrane Fusion Proteins

JUDY WHITE,[1]† ROBERT DOMS,[1] MARY-JANE GETHING,[2]‡
MARGARET KIELIAN,[1] AND ARI HELENIUS[1]

*Department of Cell Biology, Yale University School of Medicine, New Haven,
Connecticut 06510,[1] and Cold Spring Harbor Laboratory, Cold Spring Harbor,
New York 11724[2]*

Enveloped viruses introduce their genomes into cells by fusing
with cellular membranes. For each virus, fusion displays a characteristic pH profile and is mediated by a fusion protein. The best-
characterized low pH-activated fusion protein is the hemagglutinin
(HA) of influenza virus. At low pH, the HA changes conformation
and inserts into the target bilayer. We have studied the molecular
basis of the conformational change and of the interaction between
HA and the target membrane in detail. (i) Sequencing the HA gene
from an X:31 variant with a shifted pH dependence of fusion
revealed the importance of a salt link located in the trimer interface
in the pH-dependent conformational change. (ii) HAs containing
site-specific mutations in the apolar N-terminal peptide of HA2
were found to differ in both the extent and pH dependence of fusion,
further implicating this peptide as the fusion sequence. (iii) Biochemical analysis showed that the low pH-induced association between HA
and the target membrane resembles that of an integral membrane
protein. Our results conform to the model that at low pH the HA
becomes an integral component of both the viral and host-cell membranes, physically bringing them close enough together to fuse.

During the past 5 years there has been renewed interest in the general field of
virus entry. In particular, significant progress has been made in our understanding of how enveloped viruses enter and infect their host cells (for recent reviews,
see references 6, 12, and 13). Two routes of entry are used. Some viruses, such
as the paramyxovirus Sendai virus (1), introduce their nucleic acid directly into
the cytoplasm by fusing with the plasma membrane at neutral pH. Others,
including togaviruses, orthomyxoviruses, and rhabdoviruses, are first endocytosed (13) and delivered to endosomes (7); there, the mildly acidic environment triggers a genome-releasing membrane fusion reaction (20, 21). Apparently, then, all enveloped viruses utilize the same basic mechanism, membrane
fusion, to introduce their infectious material into the host cell. Viruses which
can fuse at neutral pH do so at the plasma membrane, whereas those requiring
a pH of <6 do so from within the endosomal/lysosomal compartment.

In all cases studied (for review, see reference 20), fusion is induced by a
specific viral membrane "fusion protein" (e.g., the F protein of Sendai virus [8],

†Present address: Department of Pharmacology, University of California, San Francisco, CA 94143.
‡Present address: Department of Biochemistry and Howard Hughes Medical Institute, University of
Texas Health Science Center, Dallas, TX 75235.

the *env* glycoprotein of mouse mammary tumor virus [15], the spike glycoprotein [E123] of Semliki Forest virus [10], the hemagglutinin [HA] of influenza virus [19], and the G protein of vesicular stomatitis virus [4, 16]). The purpose of this short review is to bring the reader up to date on our present understanding of how viral proteins, in particular the HA of influenza virus, mediate membrane fusion.

To fuse, two membranes must approach closely, their bilayers must coalesce, and then the coalesced membranes must separate to form one united bilayer. Although the first phase of this process, bilayer apposition, seems straightforward, it is energetically highly unfavorable. In fact, membranes experience an exponentially increasing energy barrier when they approach closer than 2.0 nm (14). Our recent work (3, 17) has given us new insights into how the HA facilitates fusion by bringing the interacting membranes into intimate contact.

The HA is a trimer which projects as a rodlike spike from the viral envelope (22). There are three major requirements for the HA to be "fusion active." First, it must be processed from its precursor form (HA0) into its mature form, in which each monomer consists of two disulfide-bonded polypeptide chains, HA1 and HA2. Second, it must be anchored into a membrane via its C-terminal transmembrane domain. Third, it must be briefly exposed to low pH (19). Several groups, including our own, have begun to probe the molecular basis of HA-mediated membrane fusion. The combined results can be summarized as follows: at the pH which triggers fusion, the HA undergoes a conformational change as evidenced by alterations in its spectral (17) and antigenic (18, 23) properties and by changes in its sensitivity to proteases (3, 17) and reducing agents (5). These data suggested that at low pH the subunits of the trimer separate from one another. Concomitantly, the previously buried apolar and highly conserved (11) N-terminal "fusion peptide" of HA2 (3; M.-J. Gething, R. Doms, D. York, and J. White, J. Cell Biol., in press) is exposed. Once exposed, the peptide inserts into the target bilayer, an interaction which resembles, in most respects, that of an integral membrane protein (3). In this manner the HA becomes tightly associated with both of the fusing membranes, anchored into one via the N terminus of HA2 and into the other via the C terminus of HA2. Since, at neutral pH, these domains are found within about 3.0 nm of one another, this dual membrane interaction may help overcome the energy barrier to close approach (14) by physically bringing the two fusing membranes together (Fig. 1).

We are currently employing genetic approaches to learn more about the pH-dependent conformational change in the HA and about the amino acid requirements of the fusion peptide. We are analyzing both naturally occurring virus variants as well as genetically engineered site-specific mutants in the fusion function.

To identify residues involved in the conformational change, we characterized a natural variant of the X:31 strain of influenza virus whose pH threshold for fusion is elevated by 0.2 pH unit. The comparative properties of the wild-type and variant viruses are given in Table 1. The results confirmed the causal relationship between the pH dependence of the conformational change of HA and its ability to bind to a target bilayer and to induce fusion (3).

Sequencing of the variant HA revealed three amino acid changes from the wild type. Two of these were in antigenic sites in HA1. The third change was

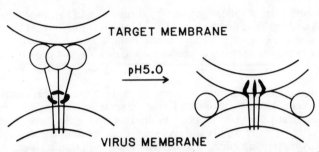

FIG. 1. Working model for how the HA of influenza virus mediates the first stage of membrane fusion: close bilayer apposition. Thickened region denotes the fusion peptide.

residue 132 of HA2 which is located in the trimer interface, near the base of the molecule (22). To determine which amino acid change was responsible for the elevated pH dependence of fusion, we constructed DNAs encoding chimeric HA molecules, expressed these in CV-1 cells using simian virus 40 late replacement vectors, and assayed them for their fusion pH threshold. The results established that the change of residue 132 in HA2 from an aspartic acid to an asparagine caused the shifted pH profile. The three-dimensional structure of the HA shows that, at neutral pH, aspartic acid 132 forms a salt link with arginine 124 of another HA2 subunit. Abolishing this salt link apparently destabilizes the interactions between HA monomers, thereby facilitating the conformational change, a prerequisite for HA-mediated membrane fusion. There are many other salt links which help stabilize the HA trimer interface (22). Interestingly, Daniels and co-workers have recently described several influenza virus mutants with elevated fusion pH optima (selected for their resistance to amantadine). The HAs from many of these mutants have amino acid substitutions which eliminate one or more of these charge interactions (2). The location of the affected salt link in our variant (~1.5 nm from the viral membrane) suggests that upon exposure to low pH, the HA trimer may dissociate all along its length.

TABLE 1. Properties of an X:31 variant with an elevated pH threshold for fusion

Property	Wild type	Variant
pK cell:cell fusion[a]	5.3	5.5
pK conformational change in HA[b]	5.4	5.7
pK BHA[c] binding to liposomes[b]	5.4	5.7
Spike morphology at pH 5.5	Rodlike	Wiry
Sensitivity to ammonium chloride (50% inhibition of infectivity)	1 mM	3 mM
Residue 132 (HA2)	Aspartic acid	Asparagine

[a] Assayed as described by White et al. (21).
[b] Assayed as described by Doms et al. (3).
[c] Bromelain-generated ectodomain of the HA.

TABLE 2. Site-specific fusion mutants of the influenza virus HA[a]

```
       1              4                      11
NH₂ gly leu phe gly ala ile ala gly phe ile glu gly gly trp gln . . .
```
$$\text{NH}_2 \text{ gly leu phe gly ala ile ala gly phe ile glu gly gly trp gln} \ldots$$

	1 ↓ glu	4 ↓ glu	11 ↓ gly

HA	Relative RBC:cell fusion activity at pH 5[b]	pH half max for RBC:cell fusion	Cell:cell fusion at pH 5[c]
Wild type	1.00	5.1	+++
Mutant 1	0.04 (±0.02)	NA[d]	−
Mutant 4	0.59 (±0.05)	5.3	+
Mutant 11	0.86 (±0.06)	5.1	−

[a] The wild type and all three mutants were positive for HA0 to HA cleavage and showed 95% erythrocyte (RBC) binding.

[b] CV-1 cells were infected with simian virus 40 HA vectors, and at 48 h postinfection they were allowed to bind RBCs which had been preloaded with horseradish peroxidase. The cells were treated for 1 min with phosphate-buffered saline buffered to pH 5 and were recultured for 1 h in normal cell culture medium; then the excess bound RBCs were removed with neuraminidase. Finally, the cell-associated horseradish peroxidase activity was enzymatically assayed. The details of this method will be published elsewhere.

[c] Assayed as described by White et al. (19).

[d] Not applicable.

To probe the amino acid requirements of the fusion peptide, we engineered specific mutations into the N terminus of HA2. We investigated the effects of introducing acidic residues within the N-terminal stretch of 10 apolar amino acids and the effect of substituting a neutral residue for the highly conserved glutamic acid found at position 11 (11). The mutant HAs were expressed in CV-1 cells and were assayed for fusion activity and for their ability to undergo the conformational change and bind to liposomes. The results are summarized in Table 2. A glycine at position 11 (instead of the conserved glutamic acid) did not alter the pH dependence or efficiency of erythrocyte fusion. However, replacement of the glycine at position 4 with a charged residue (glutamic acid) decreased the fusion activity by about 40% compared with the wild type and increased the fusion pH threshold by about 0.25 pH unit. Substitution of a glutamic acid at position 1 (for a glycine) abolished the ability of the HA to induce fusion of erythrocytes to CV-1 cells. Only the mutant 4 protein induced extensive CV-1 polykaryon formation, a phenomenon which is a more stringent measure of fusion activity (19). Somewhat surprisingly, despite these differences in fusion phenotypes, all of the mutant HA proteins changed conformation at low pH and bound tightly to liposomes. The results indicate that although a pH-dependent association of the HA2 N-terminal region is necessary for fusion, it alone is not sufficient. Therefore, in addition to mediating the first stage of fusion (bilayer apposition), the HA may also play a role in the subsequent steps of the fusion reaction. Perhaps the fusion peptide must assume a precise structure in the target membrane to cause sufficient destabilization to promote fusion.

From an examination of primary sequence data, most of the other known viral membrane fusion proteins possess putative "fusion peptides," operationally defined as stretches of uncharged amino acids ≥10 residues in length which are, in many cases, conserved within, but not between, virus families (20). Some, like the HA fusion peptide, are N terminal (e.g., the fusion peptides of the Sendai F protein and the *env* glycoprotein of mouse mammary tumor virus), while others are internal (e.g., the putative fusion peptides of Semliki Forest virus, Sindbis virus, and Rous sarcoma virus [9, 20]). Preliminary data suggest that the F protein of Sendai virus (8) and the spike glycoprotein of Semliki Forest virus (see Helenius et al., this volume) must also undergo a conformational change to be active in fusion. Although this similarity to the HA of influenza virus is tantalizing, the generality of the mechanism of HA-mediated fusion remains to be established. One challenge for the future is, therefore, to determine how each of these structurally unique fusion proteins mediates the same basic process of membrane fusion which leads to infection by most, if not all, enveloped viruses.

LITERATURE CITED

1. **Choppin, P. W., and R. W. Compans.** 1975. Replication of paramyxoviruses, p. 94–178. *In* H. Fraenkel-Conrat and R. Wagner (ed.), Comprehensive virology, vol. 4. Plenum Publishing Corp., New York.
2. **Daniels, R. S., J. C. Downie, A. J. Hay, M. Knossow, J. J. Skehel, M. L. Wang, and D. C. Wiley.** 1985. Fusion mutants of influenza virus hemagglutinin glycoprotein. Cell **40:**431–439.
3. **Doms, R., A. Helenius, and J. White.** 1985. Membrane fusion activity of the influenza virus hemagglutinin. The low pH-induced conformational change. J. Biol. Chem. **260:**2973–2981.
4. **Florkiewicz, R. Z., and J. K. Rose.** 1984. A cell line expressing vesicular stomatitis virus glycoprotein fuses at low pH. Science **225:**721–723.
5. **Graves, P. N., J. L. Schulman, J. F. Young, and P. Palese.** 1983. Preparation of influenza virus subviral particles lacking the HA1 subunit of hemagglutinin: unmasking of cross-reactive HA2 determinants. Virology **126:**106–116.
6. **Helenius, A., M. Marsh, and J. White.** 1980. The entry of viruses into animal cells. Trends Biochem. Sci. **5:**104–106.
7. **Helenius, A., I. Mellman, D. Wall, and A. Hubbard.** 1983. Endosomes. Trends Biochem. Sci. **8:**245–250.
8. **Hsu, M.-C., A. Scheid, and P. Choppin.** 1981. Activation of the Sendai virus protein (F) involves a conformational change with exposure of a new hydrophobic region. J. Biol. Chem. **256:**3557–3563.
9. **Hunter, E., E. Hill, M. Hardwick, A. Brown, D. Schwartz, and R. Tizard.** 1983. Complete sequence of the Rous sarcoma virus *env* gene: identification of structural and functional regions of its product. J. Virol. **46:**920–936.
10. **Kondor-Koch, C., B. Burke, and H. Garoff.** 1983. Expression of Semliki Forest virus proteins from cloned complementary DNA. I. The fusion activity of the spike glycoprotein. J. Cell Biol. **97:**644–651.
11. **Lamb, R. A.** 1983. The influenza virus RNA segments and their encoded proteins, p. 21–30. *In* P. Palese and D. W. Kingsbury (ed.), Genetics of influenza viruses. Springer-Verlag, New York.
12. **Lenard, J., and D. Miller.** 1982. Uncoating of enveloped viruses. Cell **28:**5–6.
13. **Marsh, M.** 1984. The entry of enveloped viruses into cells by endocytosis. Biochem. J. **218:**1–10.
14. **Rand, R.** 1981. Interacting phospholipid bilayers: measured forces and induced structural changes. Annu. Rev. Biophys. Bioeng. **10:**277–314.
15. **Redmond, S., G. Peters, and C. Dickson.** 1984. Mouse mammary tumor virus can mediate cell fusion at reduced pH. Virology **133:**393–402.
16. **Riedel, H., C. Kondor-Koch, and H. Garoff.** 1984. Cell surface expression of fusogenic vesicular stomatitis G protein from cloned cDNA. EMBO J. **3:**1477–1483.
17. **Skehel, J. J., P. M. Bayley, E. B. Brown, S. R. Martin, M. D. Waterfield, J. M. White, I. A. Wilson, and D. C. Wiley.** 1982. Changes in the conformation of influenza virus haemagglutinin at the pH optimum of virus-mediated membrane fusion. Proc. Natl. Acad. Sci. USA **79:**968–972.
18. **Webster, R., L. Brown, and D. Jackson.** 1983. Changes in the antigenicity of the hemagglutinin molecule of H3 influenza virus at acidic pH. Virology **126:**587–599.

19. **White, J., A. Helenius, and M.-J. Gething.** 1982. Haemagglutinin of influenza expressed from a cloned gene promotes membrane fusion. Nature (London) **300:**658–659.
20. **White, J., M. Kielian, and A. Helenius.** 1983. Membrane fusion proteins of enveloped animal viruses. Q. Rev. Biophys. **16:**2:151–195.
21. **White, J., K. Matlin, and A. Helenius.** 1981. Cell fusion by Semliki Forest, influenza, and vesicular stomatitis viruses. J. Cell Biol. **89:**674–679.
22. **Wilson, I., J. Skehel, and D. Wiley.** 1981. Structure of the hemagglutinin membrane glycoprotein of influenza virus at 3Å resolution. Nature (London) **289:**366–373.
23. **Yewdell, J., W. Gerhard, and T. Bachi.** 1983. Monoclonal anti-hemagglutinin antibodies detect irreversible antigenic alterations that coincide with the acid activation of influenza virus A/PR/834-mediated hemolysis. J. Virol. **48:**239–248.

Monoclonal Antibodies as Probes for Influenza Virus Hemagglutinin Structure and Function during the Infectious Cycle

JONATHAN W. YEWDELL,[1] ALEX TAYLOR,[1] ANDREW CATON,[1]
WALTER GERHARD,[1] AND THOMAS BACHI[2]

Wistar Institute for Anatomy and Biology, Philadelphia, Pennsylvania 19104,[1]
and University of Zurich, Zurich, Switzerland[2]

We have used a monoclonal antibody (Y8-10C2) specific for the influenza virus hemagglutinin (HA) to examine the structure of the HA during the infectious cycle. Under standard assay conditions this antibody binds intact virus only after acid-induced triggering of an irreversible conformational change in the HA which coincides with induction of viral fusion activity. Using immunofluorescence techniques to examine infection of MDCK cells, we found that Y8-10C2 binds internalized virus but not virus adsorbed to the cell surface and that this binding can be inhibited by ammonium chloride treatment of cells. This provides evidence that upon exposure to the acidic environment of the endosome the HA undergoes conformational changes similar to those observed in vitro. Y8-10C2 was also found to bind to newly synthesized HA, but not to mature forms of cytoplasmic HA or HA incorporated into the cell membrane. A possible common denominator of the antigenic changes detected by Y8-10C2 may be alterations in the relationship of the three monomers which form the mature HA molecule.

ANTIGENIC STRUCTURE OF THE INFLUENZA A VIRUS HA

Recognition of a protein antigen by an antibody is dependent on both the spatial relationships of the amino acid residues which form the determinant (epitope) and the accessibility of the epitope to antibody (3). In combination, these factors make antibody binding a sensitive measure of protein conformation, particularly when individual monoclonal antibodies are used as probes. Here we describe the application of a monoclonal antibody to the analysis of conformational changes occurring in the hemagglutinin (HA) of influenza virus PR8 (A/Puerto Rico/8/34). The HA is a trimeric molecule responsible for viral attachment to host cells. Antibodies specific for the HA neutralize viral infectivity, probably in most cases by blocking the attachment step. Our initial efforts focused on understanding the antigenicity of this molecule. The strategy employed took advantage of the ability of monoclonal anti-HA antibodies to select mutant viruses which escape neutralization because they fail to bind the antibody used for selection (5). Direct sequencing of genomic RNA derived from the variants revealed that their altered antigenicity could be attributed to single amino acid alterations in the HA (2). By locating these changes on the

60

three-dimensional structure of a related HA molecule (14), we found that the surface of the PR8 HA can be divided into four immunodominant regions which we termed antigenic sites Sa, Sb, Ca, and Cb (2). Sites Sa and Sb are formed by different faces of the HA tip, Cb is formed by an area near the "hinge" region, and Ca is formed by an area spanning the interface of adjacent monomers in the trimer. The binding sites of virtually all of the monoclonal anti-HA antibodies we have produced (over 200) could be at least approximately located by determining their reactivity with our defined mutant virus panel. We have used a number of these antibodies to study the HA at various stages of the influenza virus infectious cycle.

CONFORMATIONAL ALTERATIONS IN THE HA OCCURRING DURING VIRAL PENETRATION

All known functions of the HA are operative during the initial stages of the infectious cycle. In addition to its role in viral attachment, the HA is now known to mediate the fusion of viral and cellular membranes. Two conditions must be met for fusion to occur. First, the HA must be cleaved by trypsin or cellular trypsinlike proteases into disulfide-linked subunits (termed HA1 and HA2) (7, 8). This cleavage generates a hydrophobic N-terminal region on the HA2 which shows high sequence homology with an analogous polypeptide region on the paramyxovirus fusion protein (6) and which is thought to be involved in the fusion process. Second, membrane-bound virus must be exposed to mildly acidic conditions (pH 5 to 6) (9). The biological relevance of this acid-dependent fusion phenomenon is supported by recent studies which suggest that viral penetration occurs via fusion with cellular membranes following internalization and exposure to the acidic environment of the prelysosomal endosome (10, 16). Skehel and collaborators provided considerable insight into the mechanism of acid-mediated fusion by showing that irreversible conformational changes are caused in bromelain-released HA (a soluble form of HA lacking the carboxy-terminal HA2 residues that anchor the HA into the plasma membrane) by treatment with pH 5 buffer (12). Conformational changes could be detected both in the HA1 subunit, where two trypsin-sensitive sites were exposed, and in the HA2 subunit which was responsible for the acid-induced aggregation of the bromelain-released HA. This latter result was consistent with the idea that acid treatment exposes the hydrophobic HA2 amino-terminal decapeptide which could then mediate fusion of viral and cellular membranes.

The findings of Skehel et al. prompted us to examine the ability of monoclonal antibodies to detect acid-induced irreversible conformational changes in the HA1 subunit (15). Forty-two monoclonal antibodies were tested in hemagglutination inhibition (HI) assays against untreated PR8 virus and PR8 virus which had been incubated for 2 h in pH 5 buffer and returned to neutral pH for assay. The majority of the antibodies (70%) exhibited significant (fourfold or greater) changes in HI titer. With a few exceptions, the type of change (increased or decreased titer) was related to the antigenic site recognized by the antibodies. Antibodies to sites Sa and Sb had reduced HI activity against acid-treated virus, while anti-Cb antibodies had increased titers to acid-treated virus. Anti-Ca antibodies had reduced HI titers, although a few demonstrated greater HI titers.

One of the notable exceptions to this pattern was antibody Y8-10C2. This antibody clearly mapped to the Sa site by its binding to the mutant virus panel. Yet, unlike all other anti-Sa antibodies tested, it failed to detectably inhibit hemagglutination by untreated virus while efficiently inhibiting hemagglutination by acid-treated virus (HI titer of 50,000). Of all the antibodies tested in HI assays Y8-10C2 showed the greatest differences between untreated and acid-treated virus; it was therefore chosen as a probe for acid-induced antigenic alterations in subsequent experiments.

These experiments revealed that the antigenic alteration detected by Y8-10C2 exhibited a pH dependence similar to that of viral fusion activity and also coincided with the acid-induced loss of viral fusion activity and infectivity which occurs when virus is exposed to acidic conditions while not bound to cellular membranes. The ultrastructural events accompanying acid-induced conformational alterations in the HA of intact virions were followed by electron microscopy of negatively stained virus preparations. Acid-treated particles exhibited a less distinct appearance of their surface projections (HA trimers) and some slight damage to their envelope structure. Incubation of untreated and treated particles with Y8-10C2 prior to fixation and staining revealed that the antibody bound only to the surface projections present on acid-treated virions. From these observations we concluded, first, that acid-induced alterations in antigenicity are due to changes in the HA itself and are not merely a reflection of virus aggregation or disintegration, and, second, that the activity of Y8-10C2 in HI tests against acid-treated virus results from conformational alterations in the HA which lead either to the creation of a new antigenic determinant on the HA or the exposure of a previously inaccessible determinant. Although the relevance of these alterations to the fusion process was uncertain, it was clear that the binding of Y8-10C2 could serve as a useful probe for structural alterations in the HA which occur during the infectious cycle.

To this end Y8-10C2 was used in immunofluorescence studies performed on MDCK cells infected at high multiplicity with PR8 (1). Immunoreactions were performed on unfixed cells to detect HA present on the cell surface and on cells fixed with paraformaldehyde and permeabilized with Triton X-100 to detect internalized HA. In contrast to other anti-HA antibodies which detected virus-associated HA on the cell surface, Y8-10C2 detected only internalized HA, which appeared concentrated in fluorescent spots arising 10 min after infection at 37°C. The number of fluorescent spots increased steadily during the next 20 min of incubation and decreased thereafter. The fate of the HA during penetration was further investigated by infecting cells in the presence of ammonium chloride (which has been shown to neutralize the pH of the endosomal compartment [11] and to inhibit influenza virus infection [10]). This treatment resulted in the accumulation of virus in the cytoplasm (detected by other anti-HA antibodies as fluorescent spots identical in appearance to those observed with Y8-10C2 in the absence of ammonium chloride) which, however, remained nonreactive with Y8-10C2. These findings provided evidence (i) that the acid-induced conformational alterations, which coincide in vitro with the activation of viral fusion activity, also occur intracellularly and (ii) that fusion of viral and cellular membranes occurs only after internalization.

CONFORMATIONAL ALTERATIONS IN THE HA OCCURRING DURING BIOSYNTHESIS

To study conformational alterations in the HA occurring during its biosynthesis, MDCK cells were infected at a low multiplicity (at which the HA of the input virus remained undetectable during the attachment and penetration stages described above). Under these conditions, HA was first detected 90 min after infection at 37°C by both Y8-10C2 and other anti-HA antibodies as a fluorescence at the nuclear membrane extending to a small perinuclear area. At later times, Y8-10C2 fluorescence became more intense but did not occupy more cytoplasmic space. At no time in the infectious cycle did Y8-10C2 react with cell surface-associated HA, although brief treatment of intact infected cells with pH 5 buffer prior to their incubation with Y8-10C2 resulted in bright surface fluorescence. This differs from the staining observed with other monoclonal anti-HA antibodies, which spread into the cytoplasmic space and, as early as 3 h after infection, onto the cell surface. Infection of cells in the presence of tunicamycin (a potent glycosylation inhibitor) did not alter the staining pattern observed with Y8-10C2, suggesting that glycosylation is not responsible for the inability of this antibody to bind mature forms of HA. Nor could cleavage of the HA into HA1 and HA2 subunits account for the loss of Y8-10C2 reactivity since most of the HA produced during MDCK infection is uncleaved. Similar experiments performed with cells infected with a recombinant vaccinia virus containing the PR8 HA gene indicated that association of the HA with other influenza virus gene products is not necessary for the loss of Y8-10C2 reactivity. It seems most likely that the observed loss of reactivity of Y8-10C2 during HA biosynthesis is due to changes in the HA associated with trimerization. This possibility is attractive since it might explain the binding of Y8-10C2 to acid-treated virions (assuming acid treatment results in partial dissociation of monomers). This idea is supported by the findings of Daniels et al. (4) and Nestorowicz et al. (11a), who demonstrated an acid-induced dissociation of monomer-monomer contacts in the globular head of the HA and who suggested that this also accounts for the markedly decreased binding of a number of monoclonal anti-HA antibodies to acid-treated virions and isolated HA.

BINDING OF Y8-10C2 TO INTACT VIRIONS AT ELEVATED TEMPERATURES

The utility of Y8-10C2 as a probe of HA conformation in the experiments described above is in part a result of its inability to bind native HA present on intact virions under standard assay conditions (1 h at room temperature). Additional experiments have revealed, however, that binding of Y8-10C2 to virions does occur if virus and antibody are coincubated for extended times (4 to 24 h) at elevated temperatures (37°C). Binding can be detected either by electron microscopy of negatively stained virus-antibody complexes or, more conveniently, by HI or virus neutralization assays. Coincubation of virus and antibody at 37°C is essential for binding to occur; binding does not occur if virus and Y8-10C2 are preincubated separately at elevated temperatures. This indicates that binding represents detection of reversible temperature-dependent conformational alterations which occur in the HA or Y8-10C2, or both. The results of chemical cross-linking experiments suggest that antibody binding is dependent on changes in the HA. Using virus treated with various amounts of the reversible homobifunctional chemical cross-linking reagent dimethyl-3,3'-dithiobis-propionimidate (13), we found that the HI titer of Y8-10C2 following 14 h of incubation at 37°C with virus decreased with increasing amounts of cross-linker. This effect was completely reversible upon mild reduction and alkylation of cross-linked virus, which demonstrates that loss of binding is not simply due to chemical modification of the epitope recognized by Y8-10C2. Interestingly, the fusion activity of cross-linked virus (measured by hemolytic activity) paralleled Y8-10C2 HI titers. This provides further evidence for the relationship between conformational alterations responsible for Y8-10C2 binding and viral fusion activity, and suggests that the flexibility of the HA may play an important role in its ability to mediate fusion of viral and cellular membranes. Experiments are currently in progress to biochemically characterize virus treated with the minimum amount of cross-linker necessary to abrogate Y8-10C2 binding and viral hemolytic activity.

The ability of Y8-10C2 to neutralize virus after overnight incubation at 37°C has allowed us to select neutralization-resistant variants as we have previously described for other anti-HA antibodies (5). These variants fall into two classes. Class I variants represent antigenic variants which exhibit reduced reactivity with Y8-10C2 in either HI assays or indirect radioimmunoassays performed with virus bound to polyvinyl plates. Since these variants also demonstrate reduced binding with other antibodies which map to the Sa antigenic site, it is almost certain that they possess an amino acid substitution in one of the residues which constitute the Sa antigenic site. The selection of class I variants by Y8-10C2 was expected on the basis of the reduced binding of this antibody to variants selected by other anti-Sa antibodies. The selection of class II variants by Y8-10C2, on the other hand, could not have been anticipated on the basis of past experience. These variants did not demonstrate decreased binding to Y8-10C2 in indirect radioimmunoassays, nor were other antigenic alterations detected by any of the 51 other anti-HA antibodies used in this analysis. Similarly, no differences were detected by Y8-10C2 between variant and parental virus in HI assays using acid-treated virus. When HI assays were performed with untreated virus incubated with antibody for 12 h at 37°C, however, Y8-10C2 demonstrated at least a 20-fold decrease in titer with the class II variants, which presumably represents its failure to bind virus under these conditions. The simplest explanation for these findings is that Y8-10C2 fails to bind virus at 37°C not because of antigenic alterations in the Y8-10C2 epitope, but because of alterations in the HA which affect the accessibility of the epitope to antibody. Consistent with this idea are some preliminary data obtained by direct dideoxy sequencing of the genomic RNA from one of the class II variants. Sequence analysis of the entire HA1 chain revealed only a single base substitution resulting in a tyrosine-to-cysteine change at amino acid 18 (following H3 numbering). This residue is located in the stem of the HA in close proximity to the carboxy-terminal membrane-anchoring segment and in the three-dimensional structure is remote from the Sa site. While it is possible that this variant has other mutations (e.g., in HA2), the present results suggest that a mutation in the stem of the HA alters the mobility of the globular heads of the HA trimer and in this way affects recognition by Y8-10C2. Whether this mutation alone is responsible for the class II phenotype is currently under examination.

LITERATURE CITED

1. **Bachi, T., W. Gerhard, and J. W. Yewdell.** 1985. Monoclonal antibodies detect different forms of influenza virus hemagglutinin during viral penetration and biosynthesis. J. Virol. **55**:307–313.
2. **Caton, A. J., G. G. Brownlee, J. W. Yewdell, and W. Gerhard.** 1982. The antigenic structure of the influenza virus A/PR/8/34 hemagglutinin (H1 subtype). Cell **31**:417–427.
3. **Crumpton, M. J.** 1974. Protein antigens: the molecular bases of antigenicity and immunogenicity, p. 1–78. *In* M. Sela (ed.), The antigens, vol. 2. Academic Press, Inc., New York.
4. **Daniels, R. S., A. R. Douglas, J. J. Skehel, M. D. Waterfield, I. A. Wilson, and D. C. Wiley.** 1983. Studies of the influenza virus haemagglutinin in the pH5 conformation, p. 1–7. *In* W. G. Laver (ed.), The origin of pandemic influenza viruses. Elsevier Publishing Corp., New York.
5. **Gerhard, W., J. Yewdell, M. Frankel, and R. Webster.** 1981. Antigenic structure of influenza virus hemagglutinin defined by hybridoma antibodies. Nature (London) **290**:713–717.
6. **Gething, M. J., J. M. White, and M. D. Waterfield.** 1978. Purification of the fusion protein of Sendai virus: analysis of the NH$_2$-terminal sequence generated during precursor activation. Proc. Natl. Acad. Sci. USA **75**:2737–2740.
7. **Klenk, H. D., R. Rott, M. Orlich, and J. Blodorn.** 1975. Activation of influenza A viruses by trypsin treatment. Virology **68**:426–439.
8. **Lazarowitz, S. G., and P. W. Choppin.** 1975. Enhancement of infectivity of influenza A and B viruses by proteolytic cleavage of hemagglutinin polypeptide. Virology **68**:440–454.
9. **Maeda, T., and S. Ohnishi.** 1980. Activation of virus by acidic media causes hemolysis and fusion of erythrocytes. FEBS Lett. **122**:283–287.
10. **Matlin, K. S., H. Reggio, A. Helenius, and K. Simons.** 1981. Infectious entry influenza virus in a canine kidney cell line. J. Cell Biol. **91**:601–613.
11. **Maxfield, F. R.** 1982. Weak bases and inophores rapidly and reversibly raise the pH of endocytic vesicles in cultured mouse fibroblasts. J. Cell Biol. **95**:676–681.
11a.**Nestorowicz, A., G. Laver, and D. C. Jackson.** 1985. Antigenic determinants of influenza virus hemagglutinin. X. A comparison of the physical and antigenic properties of monomeric and trimeric forms. J. Gen. Virol. **66**:1687–1695.
12. **Skehel, J. J., P. M. Bayley, E. B. Brown, S. R. Martin, M. D. Waterfield, J. M. White, I. A. Wilson, and D. C. Wiley.** 1982. Changes in the conformation of influenza hemagglutinin at the pH optimum of virus-mediated membrane fusion. Proc. Natl. Acad. Sci. USA **79**:968–972.
13. **Wang, K., and F. M. Richards.** 1975. Reaction of dimethyl-3,3'-dithiobispropionimidate with intact human erythrocytes. J. Biol. Chem. **250**:6622–6626.
14. **Wiley, D. C., I. A. Wilson, and J. J. Skehel.** 1981. Structural identification of the antibody-binding sites of Hong Kong influenza hemagglutinin and their involvement in antigenic variation. Nature (London) **289**:373–378.
15. **Yewdell, J. W., W. Gerhard, and T. Bachi.** 1983. Monoclonal anti-hemagglutinin antibodies detect irreversible antigenic alterations that coincide with the acid activation of influenza virus A/PR/8/34-mediated hemolysis. J. Virol. **48**:239–248.
16. **Yoshimura, A., K. Kuroda, K. Kawasaki, S. Yamashina, T. Maeda, and S.-I. Ohnishi.** 1982. Infectious cell entry mechanism of influenza virus. J. Virol. **43**:284–293.

Membrane-Active Peptides of the Vesicular Stomatitis Virus Glycoprotein

RICHARD SCHLEGEL

Laboratory of Tumor Virus Biology, National Cancer Institute, Bethesda, Maryland 20892

Peptides corresponding to the NH_2 terminus of the vesicular stomatitis virus G protein have been synthesized and evaluated for their biological activity as well as their ability to induce neutralizing antibodies in rabbits. Antibodies specific for the G protein NH_2 terminus do not react with native protein but will recognize antigenic domains on sodium dodecyl sulfate-denatured protein. Consequently, such antibodies are nonneutralizing. The G protein peptides do exhibit several interesting biological activities; they can function efficiently as hemolysins, hemagglutinins, and cytotoxins. The hemolytic properties of some of these peptides are strongly pH dependent and occur at pH values corresponding to those observed for intact vesicular stomatitis virus or purified G protein (reconstituted into liposomes). The charge of the NH_2-terminal amino acid is crucial to the activity of these peptides. Positively charged amino acids (lysine or arginine) function well at the NH_2 terminus whereas a negatively charged amino acid (glutamic acid) abolishes peptide activity. Hemolytic peptides appear to create small membrane defects in erythrocytes, resulting in a colloid-osmotic form of hemolysis. Increased membrane permeability to ions (such as Rb^+) precedes hemoglobin release.

Infection of host cells by vesicular stomatitis virus (VSV) is a complex process involving viral attachment to the cell surface, internalization into an endosomal compartment, and finally penetration into the cell cytoplasm and uncoating (3, 4, 6, 9, 14, 16). Since there is only one glycosylated spike protein (G protein) on VSV, this protein must be multifunctional and mediate both binding and fusion with the host-cell membrane. There appears to be a limited number of high-affinity host-cell surface receptors for VSV (approximately 4,000 per Vero monkey kidney cell), as well as a large number of lower-affinity sites (15). Rabies virus, a related rhabdovirus, appears to bind to the same receptors as VSV and exhibits a similar number of saturable receptors (5,000 to 10,000) (18). Lipids appear to function as a component of the receptors for both viruses, and VSV appears to bind specifically to the acidic phospholipid phosphatidylserine (12). Removal of G protein by trypsinization dramatically reduces viral attachment and infection and confirms the role of the viral glycoprotein in attachment to the cell surface (2). After binding, VSV is rapidly translocated through plasma membrane "coated pits" into an acidic endosomal compartment. The acidic pH of this compartment is apparently a trigger for virus-cell membrane fusion, and several studies suggest that G protein directly mediates this process: (i) mammalian cells can be fused by exogenous VSV under acidic conditions (17); (ii) cells expressing cloned G protein are fusogenic at

66

low pH (7, 10); and (iii) purified G protein can effect liposome membrane fusion as detected by fluorescence energy transfer and electron microscopy (5). Purified G protein (when incorporated into liposomes) can also hemolyze erythrocytes, presumably reflecting its fusogenic function (1). In an attempt to define specific domains of the VSV glycoprotein involved in either binding or fusion, several synthetic peptides corresponding to conserved regions of the viral glycoprotein were constructed. We have generated specific antibodies to one of these peptides in an effort to interfere with specific functions of the glycoprotein and have also directly assayed the biological activities of these peptides. The biological activities of these peptides are discussed below.

SYNTHETIC PEPTIDES

Because of its highly conserved state in many strains of VSV (8, 11), its hydrophobicity, and its positive charge, we selected the NH_2 terminus of the VSV glycoprotein as a possible site involved in virus-cell membrane interaction. The synthetic peptides listed in Fig. 1 correspond exactly to the NH_2 terminus of VSV G protein or represent modifications of its structure. They were

Consensus Sequence (5 strains):

     Lys — — Ile Val Phe Pro — — — — Gly Asx Trp Lys — Val Pro — — Tyr — Tyr Cys

KFT (25) (Indiana San Juan):

     Lys Phe Thr Ile Val Phe Pro His Asn Gln Lys Gly Asn Trp Lys Asn Val Pro Ser Asn Tyr His Tyr Cys Pro

HNQ (18)

     His Asn Gln Lys Gly Asn Trp Lys Asn Val Pro Ser Asn Tyr His Tyr Cys Pro

KFT (6)

     Lys Phe Thr Ile Val Phe

RFT (6)

     Arg Phe Thr Ile Val Phe

EFT (6)

     Glu Phe Thr Ile Val Phe

KFF (6)

     Lys Phe Phe Phe Phe Phe

FIG. 1. Amino acid sequences of synthesized peptides. The consensus sequence of five VSV strains is shown along with the sequence of the NH_2 terminus of the Indiana San Juan strain. KFT(25), HNQ(18), KFT(6), RFT(6), EFT(6), and KFF(6) represent corresponding peptides (and variants) of the G protein NH_2 terminus.

synthesized and purified by high-pressure liquid chromatography by Peninsula Laboratories, Belmont, Calif., and their structure was confirmed by analysis of their amino acid composition. The sequence KFT(25) is identical with the NH_2 terminus of the mature glycoprotein of the Indiana San Juan strain of VSV, and it is clear that there is a terminal cluster of four hydrophobic amino acids at the NH_2 end of this peptide. The consensus sequence for five different strains of VSV is also shown for this region of the G protein.

ANTIBODIES TO KFT(25) REACT ONLY WITH DENATURED G PROTEIN AND DO NOT BLOCK VIRAL INFECTIVITY

Antiserum against KFT(25) was generated in rabbits and tested for its ability to recognize this domain in the native G protein. This antiserum was also tested for its ability to interfere with viral infection.

Figure 2 demonstrates that anti-KFT(25) antiserum can react with sodium dodecyl sulfate-denatured G protein in an immunoblot analysis (Fig. 2C) but that it does not react with native G protein as determined by immunofluorescence of live, infected cells (Fig. 2A and B). As might be predicted from such a finding, anti-KFT antiserum also has no effect on VSV infectivity (Table 1). Apparently the native NH_2 terminus is either inaccessible to antibodies or its conformation in the native protein is very different from that in sodium dodecyl sulfate-denatured G protein. Low pH (5.0) does not enhance the reactivity of anti-KFT(25) with G protein, suggesting that this region does not become "exposed" or assume the appropriate conformation under acidic conditions.

THE NH_2 TERMINUS OF G PROTEIN: HEMAGGLUTININ AND HEMOLYSIN

When intact VSV or purified G protein (in liposomes) is added to erythrocytes at acidic pH, hemolysis occurs. Lysis might be a consequence of fusion of damaged virions or liposomes with erythrocytes and introduction of these defects into the cell membrane. Since NH_2-terminal peptides might mimic or inhibit certain biological properties of G protein, we tested the ability of our peptides to function directly in a hemolysis assay. The results are shown in Fig. 3. Peptides corresponding to the NH_2 terminus of G protein were hemolytic at low pH. Interestingly, the most potent hemolysin, KFT(6), was a small six-amino acid peptide which exhibited half-maximal hemolysis at 3 μM concentration. A larger peptide, KFT(25), was also hemolytic but displayed half-maximal hemolysis at approximately 40 μM. When the amino-terminal lysine of KFT(6) was replaced with arginine [RFT(6)], there was little effect on the biological activity of the peptide. However, replacement of lysine with glutamic acid [EFT(6)] resulted in the total loss of hemolytic activity, suggesting that a positively charged NH_2-terminus was important for peptide-membrane interaction. Since the NH_2 end of EFT(6) is actually neutral, it is possible that a negatively charged terminus might also have functioned effectively in the peptide. Deletion of the seven NH_2-terminal amino acids from KFT(25) produces HNQ(18); this peptide lacks hemolytic properties and therefore reaffirms the role of the six NH_2-terminal amino acids in initiating cell lysis. On the basis of the above findings, we made a preliminary attempt to design a prototype

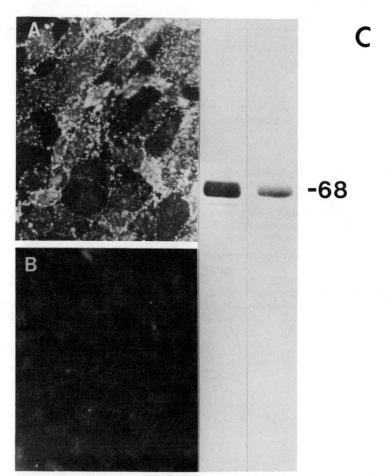

FIG. 2. Anti-KFT(25) antiserum reactivity with native and denatured G protein. (A and B) Immunofluorescence of live, VSV-infected Vero cells allowed to react with either anti-G protein antiserum (A) or anti-KFT(25) antiserum (B). Membrane fluorescence of G protein is detected only with anti-G protein antiserum. (C) Immunoblot analysis of the reaction of anti-G protein antiserum (left) and anti-KFT(25) antiserum (right) with sodium dodecyl sulfate-denatured G protein. Both antisera recognize antigenic determinants on the denatured glycoprotein.

hemolysin: a positively charged NH_2-terminal amino acid (lysine) followed by five hydrophobic amino acids (phenylalanine). However, the peptide [KFF(6)] was only hemolytic at very high concentrations (>100 to 200 μM). It is possible that our choice of a hydrophobic amino acid with a large R group might have impeded a conformation necessary for hemolysis.

TABLE 1. Lack of neutralization of VSV by anti-KFT(25) antiserum[a]

Prepn	No. of plaques
VSV	41
VSV + pre-immune serum	34
VSV + anti-KFT(25)	33
VSV + anti-G protein	0

[a] Fifty to 100 PFU of VSV were incubated with 50 μl of preimmune rabbit serum, anti-KFT(25) antiserum, or anti-G protein antiserum for 30 min at 37°C. Plaque assays were performed on Vero cells.

The same peptides which function as hemolysins are also hemagglutinins. For example, in Fig. 4, KFT(6) and KFT(25) can cause hemagglutination whereas HNQ(18) and EFT(6) cannot. It is not clear whether this reflects the ability of some of these peptides to form aggregates or whether it represents their neutralization of erythrocyte surface charge and consequent hemagglutination.

PEPTIDES INDUCE ION PERMEABILITY CHANGES BEFORE HEMOLYSIS

The mechanism by which KFT peptides cause hemolysis is unknown. However, there are indications that these peptides cause small membrane defects which result in osmotic swelling of the erythrocyte and eventually hemolysis

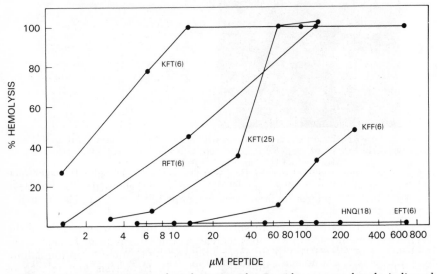

FIG. 3. Hemolytic activities of synthetic peptides. Peptides were used at the indicated concentrations, and hemolysis of erythrocytes was measured after 10 min at 37°C at pH 5.0. KFT(6) was the most potent hemolysin, with a half-maximal hemolysis at 3 μM.

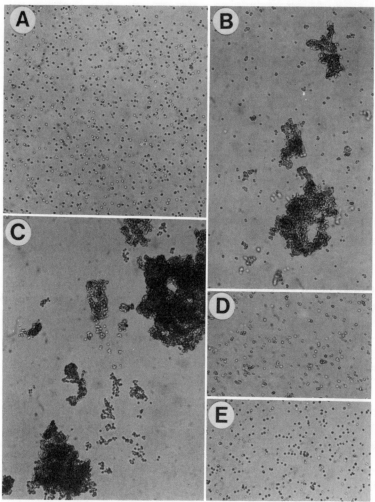

FIG. 4. Hemagglutination of sheep erythrocytes by synthetic peptides. Hemagglutination was assayed after 2 min at 37°C in the presence of (A) no peptide, (B) 50 μM KFT(25), (C) 10 μM KFT(6), (D) 200 μM HNQ(18), and (E) 200 μM EFT(6).

(13). Consistent with this hypothesis is the finding that the permeability of the erythrocyte membrane to Rb^+ ions increases dramatically prior to hemolysis. Addition of 15 μM KFT(6) to 10^9 sheep erythrocytes resulted in an immediate and rapid release of intracellular Rb^+ (Fig. 5). Hemolysis and release of hemoglobin occurred a few minutes later, presumably as the result of an accompanying influx of sodium ions and water. This dissociation of Rb^+ and

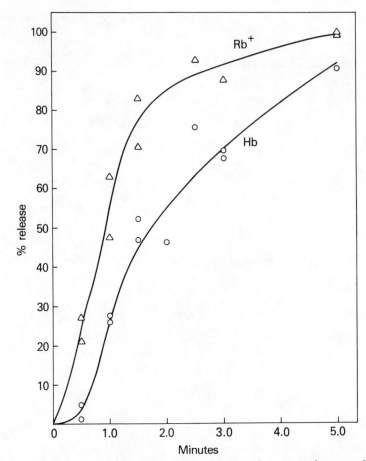

FIG. 5. Rb⁺ and hemoglobin release from sheep erythrocytes. Sheep erythrocytes (prelabeled with ^{86}Rb), 10^9, were incubated with 15 μM KFT(6) at pH 5.0 for the indicated times at 37°C. Efflux of ^{86}Rb and hemoglobin was quantitated on the same sample. Total hemolysis resulted in the release of 120,000 cpm of ^{86}Rb and 0.5 absorption units (A_{540}) of hemoglobin.

hemoglobin release from erythrocytes was exaggerated by using 3 μM KFT(6). At 1 min, 25% of Rb⁺ ion loss had occurred while none of the hemoglobin had escaped from the erythrocytes. It is unlikely that this differential release of ion and protein represents sieving of molecules through a large hole in the erythrocyte membrane. More likely, it reflects a peptide-induced change in membrane ion permeability producing a net influx of ions and water with a consequent osmotic swelling and cell lysis.

ROLE OF THE G-PROTEIN NH₂ TERMINUS IN VIRAL INFECTION

While enzyme and antibody accessibility studies suggest that the NH_2 terminus of G protein is probably not sufficiently exposed to mediate viral binding to the cell surface, it is still possible that this domain might participate in virus-cell membrane fusion. The ability of peptides from the NH_2 terminus to alter membrane permeability characteristics and initiate hemolysis suggests that there are dramatic alterations in the structure/function of the erythrocyte membrane. These ion permeability changes might reflect membrane destabilization which is presumed to occur during membrane fusion. The definition of the role of the G protein NH_2 terminus awaits genetic manipulation of the fusogenic peptide. It is very intriguing that, despite the ability to produce close apposition of cellular membranes, perturb membrane characteristics, and cause hemolysis, the NH_2-terminal G protein peptides do not produce membrane fusion. Studies on the effect of these peptides on lipid bilayer organization are currently in progress.

LITERATURE CITED

1. **Bailey, C., D. Miller, and J. Lenard.** 1984. Effects of DEAE-dextran on infection and hemolysis by VSV. Evidence that non-specific electrostatic interactions mediate effective binding of BSB to cells. Virology **133**:111–118.
2. **Bishop, D., P. Repik, J. Obijeski, N. Moore, and R. Wagner.** 1975. Restitution of infectivity to spikeless vesicular stomatitis virus by solubilized viral components. J. Virol. **16**:75–84.
3. **Dahlberg, J.** 1974. Quantitative electron microscopic analysis of the penetration of VSV into L cells. Virology **58**:250–262.
4. **Dickson, R., M. Willingham, and I. Pastan.** 1981. Alpha-2 macroglobulin adsorbed to colloidal gold: a new probe in the study of receptor-mediated endocytosis. J. Cell Biol. **89**:29–34.
5. **Eidelman, O., R. Schlegel, T. Tralka, and R. Blumenthal.** 1984. pH-dependent fusion induced by vesicular stomatitis virus glycoprotein reconstituted into phospholipid vesicles. J. Biol. Chem. **259**:4622–4628.
6. **Fan, D., and B. Sefton.** 1978. The entry into host cells of Sindbis virus, vesicular stomatitis virus, and Sendai virus. Cell **15**:985–992.
7. **Florkiewicz, R., and J. Rose.** 1984. A cell line expressing vesicular stomatitis virus glycoprotein fuses at low pH. Science **225**:721–723.
8. **Kotwal, G., J. Capone, R. Irving, S. Rhee, P. Bilan, F. Toneguzzo, T. Hofmann, and H. Ghosh.** 1983. Viral membrane glycoproteins: comparison of the amino-terminal amino acid sequences of the precursor and mature glycoproteins of three serotypes of vesicular stomatitis virus. Virology **129**:1–11.
9. **Matlin, K., H. Reggio, A. Helenius, and K. Simons.** 1982. Pathway of vesicular stomatitis virus entry leading to infection. J. Mol. Biol. **156**:609–631.
10. **Riedel, H., C. Kondor-Koch, and H. Garoff.** 1984. Cell surface expression of fusogenic vesicular stomatitis virus G protein from cloned cDNA. EMBO J. **3**:1477–1483.
11. **Rose, J., and C. Gallione.** 1981. Nucleotide sequences of mRNAs encoding the vesicular stomatitis G and M proteins determined from cDNA clones containing the complete coding regions. J. Virol. **39**:519–528.
12. **Schlegel, R., T. Tralka, M. Willingham, and I. Pastan.** 1983. Inhibition of VSV binding and infectivity by phosphatidylserine: is phosphatidylserine a VSV-binding site? Cell **32**:639–646.
13. **Schlegel, R., and M. Wade.** 1985. Biologically active peptides of the vesicular stomatitis virus glycoprotein. J. Virol. **53**:319–323.
14. **Schlegel, R., M. Willingham, and I. Pastan.** 1981. Monensin blocks endocytosis of vesicular stomatitis virus. Biochem. Biophys. Res. Commun. **102**:992–998.
15. **Schlegel, R., M. Willingham, and I. Pastan.** 1982. Saturable binding sites for vesicular stomatitis virus on the surface of Vero cells. J. Virol. **43**:871–875.
16. **Simpson, R., R. Hauser, and S. Dales.** 1969. Viropexis of vesicular stomatitis virus by L. cells. Virology **37**:285–290.
17. **White, J., K. Matlin, and A. Helenius.** 1981. Cell fusion by Semliki Forest, influenza, and vesicular stomatitis virus. J. Cell Biol. **89**:674–679.
18. **Wunner, W., K. Reagan, and H. Koprowski.** 1984. Characterization of saturable binding sites for rabies virus. J. Virol. **50**:691–697.

Identification and Analysis of Biologically Active Sites of Herpes Simplex Virus Glycoprotein D

ROSELYN J. EISENBERG[1,2] AND GARY H. COHEN[2,3]

Department of Pathobiology, School of Veterinary Medicine,[1] and Department of Microbiology[3] and Center for Oral Health Research,[2] School of Dental Medicine, University of Pennsylvania, Philadelphia, Pennsylvania 19104

Herpes simplex virus (HSV) causes human diseases including cold sores, eye and genital infections, and encephalitis. Glycoprotein D (gD) is a virion envelope component of the oral (HSV-1) and genital (HSV-2) forms. Evidence to date suggests that gD functions in virus absorption and fusion of infected cells and is a major target of the immune response. Our goal is to relate biological functions of gD to specific portions of the glycoprotein. Using a panel of gD-specific monoclonal antibodies, we have defined eight antigenic sites. Four discontinuous epitopes have been partially localized, and their relative positions have been determined by competition analysis. Three continuous epitopes were localized, and synthetic peptides mimicking their sequences reacted with the monoclonal antibodies. The relationship of these epitopes to induction of neutralizing antibody and protection was assessed. Synthetic peptides mimicking the first 23 amino acids of gD induced type-common virus-neutralizing antibody. Mice immunized with peptides were protected from virus challenge by either the intraperitoneal or footpad route. Thus, the amino terminus of gD is biologically active and may constitute an important component of a subunit vaccine.

Herpes simplex viruses (HSVs) cause a number of human diseases, including cold sores, eye and genital infections, and encephalitis (30). Glycoprotein D (gD) of HSV is a structural component of the virion envelope which stimulates production of high titers of virus-neutralizing activity (6–8, 10, 15, 17, 27) and is likely to play an important role in the initial stages of viral infection. It was recently shown that certain anti-gD monoclonal antibodies (MCAb) block fusion of infected cells (31), whereas others block virus absorption (19). In addition, animals immunized with gD are protected from an HSV challenge (2, 3, 11, 25, 27, 32). Tryptic peptide analysis (1, 17) and amino-terminal sequencing (14) showed that gD of HSV type 2 (HSV-2) (gD-2) is structurally similar though not identical to gD of HSV-1 (gD-1). Recently, the genes for gD-1 and gD-2 were localized and sequenced (24, 26, 38, 39). Although the deduced amino acid sequences for the two glycoproteins were shown to be 85% homologous, little is known about their secondary or tertiary structure. Using MCAb, we previously defined eight epitopes within gD (15, 16) some of which are type common and others of which are type specific. We have now analyzed a number of additional anti-gD MCAb, and most fit into our original groupings. The high degree of amino acid sequence homology between gD-1 and gD-2 probably accounts for

74

the immunological cross-reactivity of polyclonal antibodies and MCAb directed against gD (15, 33). On the other hand, the type specificity of other MCAb is undoubtedly related to differences in the structures of gD-1 and gD-2 and, consequently, in amino acid sequence. Recently, two type 2-specific antigenic sites were localized to the first 23 amino acids of gD (12). The type specificity was shown to be due to the two amino acid differences between gD-1 and gD-2 at positions 7 and 21.

Our goal is to relate the structure of the protein to its biological functions. We have used a number of approaches to localize and characterize the specific epitopes of gD which react with each of the MCAb groups. In addition, a number of investigators have examined some of the biological properties of the MCAb (1, 13, 15, 16, 19, 31, 33, 37). The results so far indicate that different parts of the glycoprotein can be related to specific biological functions. For example, we used synthetic peptides to show that the type-common group VII epitope was localized to residues 11 to 19 of the mature form of gD (6, 7, 12). Polyclonal sera to certain of the synthetic peptides in the region of the first 23 amino acids of gD-1 and gD-2 exhibited type-common virus-neutralizing activity. Mice immunized with these peptides were protected from a lethal HSV-2 challenge (6). Thus the first 23 amino acids of gD represent a biologically important part of the glycoprotein.

We have continued to delineate the location and characteristics of the antigenic epitopes of gD. First, we distinguished between discontinuous and continuous epitopes, i.e., those which are lost under the denaturing conditions of reduction and alkylation (boiling of the protein in the presence of sodium dodecyl sulfate plus mercaptoethanol followed by iodoacetamide treatment) from those which are retained (16). In addition to group VII, groups II and V recognize continuous epitopes of gD. These have been specifically localized to residues 268–287 and 340–356, respectively (16). In addition, we have analyzed fragments of gD and carried out competition studies to map the relative positions of four discontinuous epitopes, corresponding to MCAb in groups I, III, IV, and VI (16).

PREDICTED SECONDARY STRUCTURES OF gD-1 AND gD-2

Figure 1 shows a computer representation (7) of the secondary structure of gD-1 (Fig. 1A) and gD-2 (Fig. 1B), derived from the predicted amino acid sequences (24, 38, 39) and rules established by Chou and Fasman (4, 5). We have also analyzed both glycoproteins with a second empirical analysis (Fig. 1) which assumes that hydrophilic regions of protein structure have a greater immunological potential (22). For these calculations, the first 25 amino acids of the predicted sequence were excluded from consideration, since direct N-terminal sequence analysis showed that, for both gD-1 and gD-2, lysine residue 26 of the deduced sequence was the amino terminus of the mature protein (14). In our numbering system this lysine is amino acid residue 1. Our working hypothesis has been that epitopes are likely to be located in regions where highly hydrophilic residues are present in beta turns (36). If, in addition, the epitope is continuous, synthetic peptides could be used to mimic the reactivity of the epitope. The positions of the predicted N-asparagine-linked carbohydrates

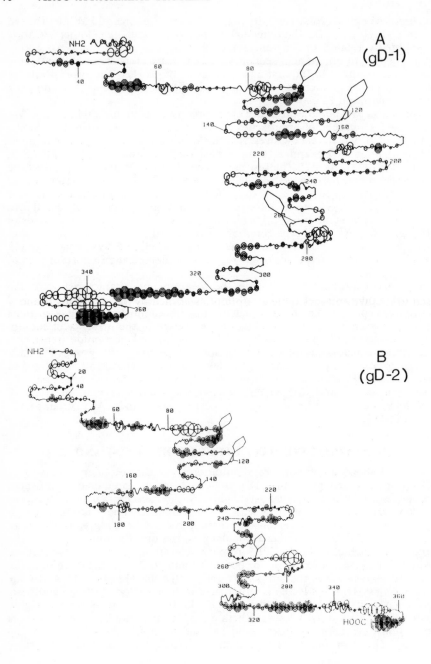

(shown as balloons) are based on the sequence Asn-X-Thr or Asn-X-Ser (23). For gD-1 and gD-2, all three positions are glycosylated (9). The homology in amino acid sequence is reflected in similarities in both secondary structure and regions of hydrophilicity in the two proteins. In at least one case, however, two differences in amino acid sequence in region 1 to 23 (24, 38, 39) have been correlated with both changes in predicted secondary structure and antigenicity (7, 12). For both gD-1 and gD-2, there are two regions in which beta turns intersect a highly hydrophilic region, i.e., residues 11–19 and 268–287. A third region in gD-1, residues 340–356, is hydrophilic and contains a predicted beta turn. In gD-2, however, no beta turn is predicted in this region.

REACTION OF MCAb AGAINST NATIVE AND DENATURED gD

We used the immunoblot assay to examine the nature of the epitope as well as the location of the epitope specified by different groups of MCAb. Figure 2A shows that groups I, II, III, V, and VII react with gD-1 and gD-2 (rows 2, 3, 6, and 8, lanes a and b) whereas groups IV and VI are type specific (rows 5 and 7, lanes a and b). In addition, three groups of MCAb, groups II, V, and VII, reacted with both native and denatured gD (Fig. 2A, rows 3, 6, and 8, lanes a through d) and four groups, I, III, IV, and VI, reacted only with native gD (rows 2, 4, 5, and 7, lanes a through d). A truncated form of gD-1, containing residues 1–275 was used to localize epitopes (lane e). Epitopes specified by MCAb groups II and V are located downstream of residue 275, since these antibodies failed to react with the truncated molecule. The other five epitopes are located upstream of residue 275.

LOCATION OF THE GROUP II AND V EPITOPES

Previous studies employing an immunofluorescence assay showed that group V MCAb reacted with fixed, but not with unfixed, HSV-1-infected cells (29). This suggested that the epitope was not exposed on the external face of the plasma membrane of infected cells. Moreover, when gD-1 was synthesized and processed in an in vitro system, the processed protein was partially protected from proteolysis by trypsin (29). Approximately 3,000 daltons of the protein was removed by this treatment, and the trypsin-resistant fragment could not be immunoprecipitated by group V MCAb. Furthermore, when truncated forms of the gD gene, lacking the information for the transmembrane-anchoring region plus the carboxy terminus, were cloned into *Escherichia coli*, the expressed

FIG. 1. Predicted secondary structure and hydrophilicity maps of gD-1 (A) and gD-2(B). Secondary structures were predicted by a computer program, using the rules of Chou and Fasman for determining Pt, Pa, and Pb (4, 5, 16). Shaded circles indicate hydrophobic regions; open circles indicate hydrophilic areas. The radius of a circle over a residue is proportional to the mean hydrophilicity as calculated for that residue plus the next five residues (22). The value is therefore distorted at the C-terminal end. The hexagonal balloons indicate predicted sites (Asn-X-Thr or Ser) of N-asparagine-linked glycosylation (23).

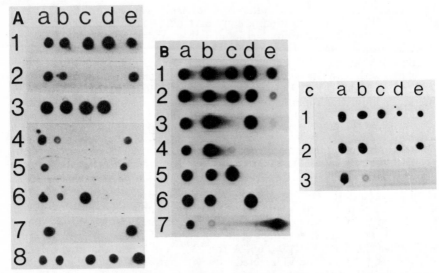

FIG. 2. Immunoblot analyses of (A) MCAb directed at HSV gD; (B) synthetic peptides which mimic portions of gD-1, using polyclonal antibodies and MCAb; and (C) truncated forms of gD-1, using MCAb. (A) The antibodies used (15) were: row 1, anti-gD-1 (rabbit 1); row 2, group I, HD-1; row 3, group II, DL6; row 4, group III, 11S; row 5, group IV, 41S; row 6, group V, 57S; row 7, group VI, 45S; row 8, group VII, 170. Antigens: Lane a, immunosorbent-purified (18) native gD-1 (15 ng); lane b, native gD-2 (15 ng); lane c, denatured gD-1 (15 ng); lane d, denatured gD-2 (15 ng); lane e, truncated gD-1, residues 1 to 275 (12) (60 ng). (B) Immunoblot analysis of synthetic peptides which mimic portions of gD-1, using polyclonal antibodies and MCAb. The antibodies used were: row 1, anti-gD-1 (rabbit 1); row 2, anti-gD-1 (rabbit 2); row 3, anti-gD-2 (rabbit 3); row 4, anti-gD-2 (rabbit 4); row 5, group VII, 170; row 6, group II, DL6; row 7, group V, 57S. Antigens: lane a, native gD-1 (15 ng); lane b, native gD-2 (15 ng); lane c, peptide 8–23[1] (7, 24, 25) (500 ng); lane d, 268–287 (1 μg); lane e, 340–356 (100 ng). Peptide 8–23[1] was synthesized by procedures described by Cohen et al. (7). Peptides 268–287 and 340–356 were purchased from Peninsula Laboratories, Inc. (C) Immunoblot analysis of truncated forms of gD-1, using MCAb. The antibodies used were: row 1, group VII, 170; row 2, group II, DL6; row 3, group V, 57S. Antigens: lane a, native gD-1 (15 ng); lane b, native gD-2 (15 ng); lane c, truncated gD-1 residues 1 to 275 (25) (60 ng); lane d, truncated gD-1, residues 1 to 287 (16, 20; M. G. Gibson and P. G. Spear, J. Cell Biochem. Suppl. 8B, abstr. no. 1337, p. 191, 1984), 100 ng of protein; lane e, truncated gD-1, residues 1 to 287, 200 ng of protein.

gD-like protein was not recognized by 57S, a group V antibody (40). These results, taken together, suggested that the group V epitope was located between residues 340 and 369 of gD-1.

To localize the epitope further, we relied on computer predictions (Fig. 1; 4, 5, 16, 22) as well as differences in the sequences of gD-1 and gD-2 at the carboxy terminus (24, 38, 39). The epitope appeared to be type 1 specific by immuno-precipitation analysis (15), but similarities in the sequence lead to some cross-reactivity detectable by immunoblot (Fig. 2A; reference 16). The region from 340 to 356 showed the most differences in both primary amino acid sequence and predicted secondary structure in the two glycoproteins. Moreover, in gD-1 this region contained hydrophilic residues within a beta turn. Therefore, a synthetic peptide, mimicking this sequence in gD-1, was tested against group V MCAb (Fig. 2B). It is clear that this peptide reacts with group V MCAb (Fig. 2b, row 7, lane e) and with three polyclonal anti-gD antibodies (rows 1, 2, and 3, lane e), but not with other anti-gD MCAb.

Preliminary localization of the group II epitope was accomplished by testing the reactivity of two truncated forms of gD-1, representing residues 1–275 (16, 25) and 1–287 (16, 20) against DL6 MCAb (Fig. 2C). Both truncated forms reacted with group VII antibody (Fig. 2C, row 1, lanes c, d, and e), indicating the presence of residues 8–23 in the truncated proteins. Neither form reacted with group V antibody (Fig. 2C, row 3, lanes c, d, and e). The truncated form 1–275 also failed to react with group II MCAb (Fig. 2C, row 2, lane c), but the truncated form 1–287 did react (row 2, lanes d and e). These results suggested that the group II epitope was probably located between residues 275 and 287, although it could include several upstream residues.

On the basis of these data, as well as computer predictions that the region from 266 to 282 was highly hydrophilic and contained a beta turn (4, 5, 16, 22), we tested the reactivity of a synthetic peptide, mimicking residues 268–287 of gD-1 (268–287) (Fig. 2B). The peptide reacted with group II MCAb (Fig. 2B, row 6, lane d), but not with antibodies in group V (row 7, lane d) or VII (row 5, lane d). In addition, the peptide reacted against three of four polyclonal sera prepared against gD (Fig. 2B, rows 1 to 3, lane d).

FUNCTIONS OF THREE CONTINUOUS EPITOPES OF gD

Table 1 summarizes current information concerning the biochemical and biological properties of the three continuous epitopes specified by MCAb groups VII, II, and V. The epitope specified by group V MCAb is not expressed on the surface of infected cells. This MCAb does not exhibit any neutralizing activity and does not protect mice in passive immunization studies (35). Antisera prepared to the synthetic peptide mimic the activity of the group V MCAb. The epitopes specified by MCAb groups II and VII are each expressed on the surface of infected cells, and the corresponding MCAb exhibit type-common virus-neutralizing activity. In addition, the MCAb have been shown to protect mice in passive immunization studies (R. Dix, personal communication). Also, the synthetic peptide mimicking residues 1–23, and containing the group VII epitope, when coupled to keyhole limpet hemocyanin, stimulated production of neutralizing antibody that reacted with gD from infected cells (6, 7, 12). Mice immunized with this peptide were protected from a lethal virus challenge (6). However, the coupled peptide for group II (residues 268–287) failed to stimulate production of neutralizing antibody (16). The reason for this lack of im-munogenicity is not clear, since several polyclonal sera prepared against gD contain antibody that reacts with the synthetic peptide (Fig. 2B) and since group II MCAb neutralize virus. The data so far indicate that the group II and VII epitopes are biologically important in terms of protective immunity.

TABLE 1. Properties of continuous epitopes of HSV gD

Property	Monoclonal antibody group			Reference
	VII	II	V	
Residues in gD	11–19	268–287	340–356	6, 7, 16
Epitope on infected cell surface	+	+	−	24
Antibody reacts with gD-1	+	+	+	15, 37
Antibody reacts with gD-2	+	+	+/−	15, 37
Antibody neutralizes HSV	+	+	−	16, 37
Antibody protects in passive immunization	+++	+	−	35; Dix, personal communication
Peptide stimulates antibody to gD	+	−	+	6, 7, 12, 16
Peptide stimulates neutralizing antibody	+	−	−	6, 7, 16
Peptide protects mice against HSV	+	ND[a]	ND	6

[a] ND, Not done.

LOCALIZATION AND PROPERTIES OF DISCONTINUOUS EPITOPES OF gD

Figure 2A (lane e) shows that group I, III, IV, VI, and VII MCAb bound to the truncated form of gD 1–275. These results together with those of previous V8 protease experiments (15, 16) indicate that all of the discontinuous epitopes are localized in an amino-terminal fragment (1–256) which excludes one glycosylation site as well as the transmembrane anchor region.

Table 2 summarizes current information concerning the biochemical and biological properties of epitopes in groups I, III, IV, and VI. Group I MCAb exhibit very high titers of type-common virus-neutralizing activity which is independent of complement (15, 16, 37). The epitope appears to be highly conserved, possibly identical, in gD-1 and gD-2 since group I MCAb reacts equally well with all strains of HSV-1 and HSV-2 tested (33). The epitope which reacts with group I MCAb may be involved in fusion since group I MCAb inhibit this activity to the greatest extent (19. 31). Antibodies in this group have been shown to protect mice against a lethal HSV challenge in passive immunization studies (13; R. Dix, personal communication). Group III MCAb are type common, but are much more reactive against gD-1 than against gD-2 (15,16; Fig. 2A, row 4, lanes a and b). This antibody exhibits complement-dependent type-common neutralizing activity (15, 37). MCAb in groups IV and VI are type specific (15, 37). Interestingly, group VI MCAb, which exhibit either very low or no titers of virus-neutralizing activity (15, 16, 37), inhibit virus absorption to the greatest extent among gD MCAb tested so far (19). Antibodies in this group do exhibit some capacity for protection in passive immunization experiments, but only against an HSV-1 challenge (R. Dix, personal communication).

TOPOGRAPHICAL RELATIONSHIP OF DISCONTINUOUS EPITOPES

Competition studies were carried out to further define the relationship of the discontinuous epitopes to one another (16). In these experiments, gD was first

TABLE 2. Properties of discontinuous epitopes of HSV gD

Property	Monoclonal antibody group				Reference
	I	III	IV	VI	
Reacts with gD-1	+	+	+	+	15, 37
Reacts with gD-2	+	+	−	−	15, 37
Residue number in gD	1–275	1–275	1–275	1–275	15, 16, 25
Epitope on infected cell surface	+	+	+	+	
Neutralizes HSV	+++	+[a]	+	−	15, 31, 33, 37
Antibody protects in passive immunity	+++	ND[b]	ND	+	35; Dix, personal communication
Antibody inhibits fusion[c]	S[d]	I	W	W	19, 31
Antibody inhibits binding[c]	W	I	W	S	19

[a] Neutralization by group III antibodies is strongly complement dependent (19, 31, 37).

[b] ND, Not done.

[c] Antibodies kindly supplied by P. G. Spear were grouped by reactivity against gD-1 and gD-2, as well as competition studies (31; G. H. Cohen and R. J. Eisenberg, unpublished data). These antibodies had previously been analyzed for ability to inhibit fusion (31) and adsorption (19). The groupings were as follows: antibodies III-114-4 and II-174-1, group I; antibody I-99-1, group III; antibody II-436-1, group IV; antibodies I-206-7 and II-886-1, group VI.

[d] Ability to inhibit fusion or adsorption was measured (19, 31), and the levels are indicated as S, strongest; I, intermediate; and W, weakest.

allowed to react with an unlabeled MCAb and then with a second iodinated MCAb. In some cases, the binding of one antibody to gD-1 had no effect (no competition) on the binding of another. These results were further evidence that the MCAb groupings are valid and that there are distinct discontinuous epitopes on gD-1. In other cases, there was competition. This indicated that (i) some amino acids in these epitopes are shared, (ii) the epitopes were so close that there was steric hindrance in the binding of a second antibody, or (iii) binding of one antibody altered the conformation of gD so that binding of the second antibody was affected (16, 28).

On the basis of these studies, as well as the studies of the discontinuous epitopes, we have constructed a two-part topographic map for gD-1 (Fig. 3). First, we have depicted the protein essentially as a linear molecule with the positions of the three continuous epitopes indicated. The discontinuous epitopes have been depicted in a separate drawing as ellipses located downstream from group VII, each of which includes amino acids prior to residue 260. The discontinuous epitopes corresponding to MCAb which exhibited competition are shown as overlapping. Thus group III overlaps groups IV and VI and group I also overlaps groups IV and VI. Groups I and III do not overlap at all. Preliminary evidence indicates that antibodies in all of the MCAb groups except possibly group IV are able to immunoprecipitate the gD-like protein produced from tunicamycin-treated HSV-1-infected cells (16, 34). Three of the epitopes are depicted as involving disulfide bonds (S–S in Fig. 3) although we do not know how many cysteines are disulfide bonded in gD or how many are involved in determining the structure of any one epitope.

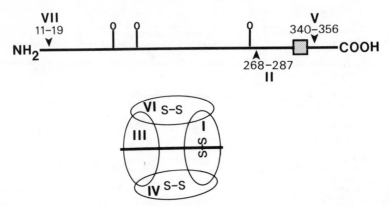

FIG. 3. Topographic map of HSV-1 gD. The positions of epitopes binding to MCAb in groups VII, II, and V are shown. Also indicated (as balloons) are the three N-asparagine-linked glycosylation sites. The transmembrane region is depicted as a box. The positions of discontinuous epitopes (ellipses at bottom) were derived from competition experiments (16). At least three of the epitopes involve disulfide bonds.

Further localization of discontinuous epitopes will require other approaches, including a more complete understanding of the contributions of disulfide bonds to the structure of gD. One possible approach will be to analyze the amino acid changes associated with mutants which exhibit an altered pattern of reactivity with MCAb. Such mutants would include those which are no longer neutralized by antibody, such as the "mar" mutants (21). Another approach which we are now taking is to examine further truncated forms of gD for reactivity against the MCAb groups and to create specific mutations in the gD gene by the techniques of site-directed mutagenesis. These approaches should allow us to localize discontinuous epitopes more precisely and to begin to understand the tertiary structure of gD. It is clear, however, that precise localization of discontinuous epitopes requires a knowledge of the complete three-dimensional structure of gD.

This work was supported by Public Health Service grants AI-18289 from the National Institute of Allergy and Infectious Diseases and DE-02623 from the National Institute of Dental Research and by a grant to R.J.E. and G.H.C. from American Cyanamid Co.

LITERATURE CITED

1. **Balachandran, N., D. Harnish, W. E. Rawls, and S. Bacchetti.** 1982. Glycoproteins of herpes simplex virus type 2 as defined by monoclonal antibodies. J. Virol. **44**:344–355.
2. **Berman, P. W., T. Gregory, D. Crase, and L. A. Lasky.** 1985. Protection from genital herpes simplex virus type 2 infection by vaccination with cloned type 1 glycoprotein D. Science **227**:1490–1492.
3. **Chan, W.** 1983. Protective immunization of mice with specific HSV-1 glycoproteins. Immunology **49**:343–352.
4. **Chou, P. Y., and G. D. Fasman.** 1974. Conformational parameters for amino acids in helical beta-sheet and random coil regions. Biochemistry **13**:211–222.
5. **Chou, P. Y., and G. D. Fasman.** 1974. Prediction of protein conformation. Biochemistry **13**:222–245.

6. **Cohen, G. H., B. Dietzschold, D. Long, M. Ponce de Leon, E. Golub, and R. J. Eisenberg.** 1985. Localization and synthesis of an antigenic determinant of herpes simplex virus glycoprotein D that stimulates neutralizing antibody and protects mice from lethal virus challenge, p. 24–30. *In* S. E. Mergenhagen and B. Rosan (ed.), Molecular basis of oral microbial adhesion. American Society for Microbiology, Washington, D.C.

7. **Cohen, G. H., B. Dietzschold, M. Ponce de Leon, D. Long, E. Golub, A. Varrichio, L. Pereira, and R. J. Eisenberg.** 1984. Localization and synthesis of an antigenic determinant of herpes simplex virus glycoprotein D that stimulates the production of neutralizing antibody. J. Virol. **49:**102–108.

8. **Cohen, G. H., M. Katze, C. Hydrean-Stern, and R. J. Eisenberg.** 1978. Type-common CP-1 antigen of herpes simplex virus is associated with a 59,000 molecular weight envelope glycoprotein. J. Virol. **47:**172–181.

9. **Cohen, G. H., D. Long, J. T. Matthews, M. May, and R. J. Eisenberg.** 1983. Glycopeptides of the type-common glycoprotein gD of herpes simplex virus types 1 and 2. J. Virol. **46:**679–689.

10. **Cohen, G. H., M. Ponce de Leon, and C. Nichols.** 1972. Isolation of a herpes simplex virus-specific antigenic fraction which stimulates the production of neutralizing antibody. J. Virol. **10:**1021–1030.

11. **Cremer, K. J., M. Macket, C. Wohlenberg, A. L. Notkins, and B. Moss.** 1985. Vaccinia virus recombinant expressing herpes simplex virus type 1 glycoprotein D prevents latent herpes in mice. Science **228:**737–740.

12. **Dietzschold, B., R. J. Eisenberg, M. Ponce de Leon, E. Golub, F. Hudecz, A. Varrichio, and G. H. Cohen.** 1984. Fine structure analysis of type-specific and type-common antigenic sites of herpes simplex virus glycoprotein D. J. Virol. **52:**431–435.

13. **Dix, R. D., L. Pereira, and J. R. Baringer.** 1981. Use of monoclonal antibody directed against herpes simplex virus glycoproteins to protect mice against acute virus-induced neurological disease. Infect. Immun. **34:**192–199.

14. **Eisenberg, R. J., D. Long, R. Hogue-Angeletti, and G. H. Cohen.** 1984. Amino-terminal sequence of glycoprotein D of herpes simplex virus types 1 and 2. J. Virol. **49:**265–268.

15. **Eisenberg, R. J., D. Long, L. Pereira, B. Hampar, M. Zweig, and G. H. Cohen.** 1982. Effect of monoclonal antibodies on limited proteolysis of native glycoprotein gD of herpes simplex virus types 1 and 2 by use of monoclonal antibody. J. Virol. **41:**478–488.

16. **Eisenberg, R. J., D. Long, M. Ponce de Leon, J. T. Matthews, P. G. Spear, M. G. Gibson, L. A. Lasky, P. Berman, E. Golub, and G. H. Cohen.** 1985. Localization of epitopes of herpes simplex virus type 1 glycoprotein D. J. Virol. **53:**634–644.

17. **Eisenberg, R. J., M. Ponce de Leon, and G. H. Cohen.** 1980. Comparative structural analysis of glycoprotein gD of herpes simplex virus types 1 and 2. J. Virol. **35:**428–435.

18. **Eisenberg, R. J., M. Ponce de Leon, L. Pereira, D. Long, and G. H. Cohen.** 1982. Purification of glycoprotein gD of herpes simplex virus types 1 and 2 by use of monoclonal antibody. J. Virol. **41:**1099–1104.

19. **Fuller, A. O., and P. G. Spear.** 1985. Specificities of monoclonal and polyclonal antibodies that inhibit adsorption of herpes simplex virus to cells and lack of inhibition by potent neutralizing antibodies. J. Virol. **55:**475–482.

20. **Gibson, M. G., and P. G. Spear.** 1983. Insertion mutants of herpes simplex virus have a duplication of the glycoprotein D gene and express two different forms of glycoprotein D. J. Virol. **48:**396–404.

21. **Holland, T. C., S. D. Marlin, M. Levine, and J. Glorioso.** 1983. Antigenic variants of herpes simplex virus selected with glycoprotein-specific monoclonal antibodies. J. Virol. **45:**672–682.

22. **Hopp, T. P., and K. R. Woods.** 1981. Prediction of protein antigenic determinants from amino acid sequences. Proc. Natl. Acad. Sci. USA **78:**3824–3828.

23. **Hubbard, S. D., and R. J. Ivatt.** 1981. Synthesis and processing of asparagine-linked oligosaccharides. Annu. Rev. Biochem. **50:**555–583.

24. **Lasky, L. A., and D. Dowbenko.** 1984. DNA sequence analysis of the type-common glycoprotein D genes of herpes simplex virus types 1 and 2. DNA **3:**23–29.

25. **Lasky, L. A., D. Dowbenko, C. C. Simonsen, and P. W. Berman.** 1984. Protection of mice from lethal herpes simplex virus infection by vaccination with a secreted form of cloned glycoprotein D. Biotechnology **2:**527–532.

26. **Lee, G. T.-Y., M. F. Para, and P. G. Spear.** 1982. Location of the structural genes for glycoprotein gD and for other polypeptides in the S component of herpes simplex virus type 1 DNA. J. Virol. **43:**41–49.

27. **Long, D., T. J. Madara, M. Ponce de Leon, G. H. Cohen, P. C. Montgomery, and R. J. Eisenberg.** 1984. Glycoprotein D protects mice against lethal challenge with herpes simplex virus types 1 and 2. Infect. Immun. **43:**761–764.

28. **Lubeck, M., and W. Gerhard.** 1982. Conformational changes at topologically distinct antigenic sites on the influenza A/PR/8/34 virus HA molecule are induced by the binding of monoclonal antibodies. Virology **118**:1–7.

29. **Matthews, J. T., G. H. Cohen, and R. J. Eisenberg.** 1983. Synthesis and processing of glycoprotein D of herpes simplex virus types 1 and 2 in an in vitro system. J. Virol. **48**:521–533.

30. **Nahmias, A. J., J. Dannenbarger, C. Wickliffe, and J. Muther.** 1980. Clinical aspects of infection with herpes simplex viruses 1 and 2, p. 2–9. *In* A. J. Nahmias, W. R. Dowdle, and R. F. Schinazi (ed.), The human herpes viruses, an interdisciplinary perspective. Elsevier/North-Holland Publishing Co., New York.

31. **Noble, A. G., G. T.-Y. Lee, R. Sprague, M. L. Parish, and P. G. Spear.** 1983. Anti-gD monoclonal antibodies inhibit cell fusion induced by herpes simplex virus type 1. Virology **129**:218–224.

32. **Paoletti, E., B. R. Lipinskas, C. Samsonoff, S. Mercer, and D. Panicali.** 1984. Construction of live vaccines using genetically engineered poxviruses: biological activity of vaccinia virus recombinants expressing the hepatitis B virus surface antigen and the herpes simplex virus glycoprotein D. Proc. Natl. Acad. Sci. USA **81**:193–197.

33. **Pereira, L., D. V. Dondero, D. Gallo, V. Devlin, and J. D. Woodie.** 1982. Serological analysis of herpes simplex virus types 1 and 2 with monoclonal antibodies. Infect. Immun. **35**:363–367.

34. **Pizer, L. I., G. H. Cohen, and R. J. Eisenberg.** 1980. Effect of tunicamycin on herpes simplex virus glycoproteins and infectious virus production. J. Virol. **34**:142–153.

35. **Rector, J. T., R. N. Lausch, and J. E. Oakes.** 1984. Identification of infected cell-specific monoclonal antibodies and their role in host resistance to ocular herpes simplex virus type 1 infection. J. Gen. Virol. **65**:657–661.

36. **Rose, G. D.** 1978. Prediction of chain turns in globular proteins on a hydrophobic basis. Nature (London) **272**:586–590.

37. **Showalter, S. D., M. Zweig, and B. Hampar.** 1981. Monoclonal antibodies to herpes simplex virus type 1 proteins, including the immediate-early protein ICP 4. Infect. Immun. **34**:684–692.

38. **Watson, R. J.** 1983. DNA sequence of the herpes simplex virus type 2 glycoprotein D gene. Gene **26**:307–312.

39. **R. J. Watson, J. H. Weis, J. S. Salstrom, and L. W. Enquist.** 1982. Herpes simplex virus type 1 glycoprotein D gene: nucleotide sequence and expression in *Escherichia coli*. Science **218**:381–383.

40. **Watson, R. J., J. H. Weis, J. H. Salstrom, and L. W. Enquist.** 1985. Bacterial synthesis of herpes simplex virus type 1 and 2 glycoprotein D antigens. J. Invest. Dermatol. **83**(Suppl.):102S–111S.

Protein-Mediated Fusion of Viral and Cellular Membranes

FRANK R. LANDSBERGER AND PRAVINKUMAR B. SEHGAL

The Rockefeller University, New York, New York 10021

Influenza viruses A, B, and C and vesicular stomatitis virus each possess specific viral glycoproteins that mediate fusion of the viral membrane with a host-cell membrane at pH 5. This results in liberation of the viral genome into the cytosol. Using a new scoring method to assess conservation of amino acid sequences, we have found significant alignment of residues in the specific viral fusion proteins and the fatty acid-binding site of soluble phosphatidyl choline transport protein. This suggests that fusion by the viral proteins may involve lipid transport, leading to destabilization of the target membrane.

The entry of enveloped viruses into mammalian cells (6, 17) represents an important example of specific, protein-mediated membrane fusion. Enveloped viruses possess a lipid bilayer with which is associated a specific viral glycoprotein that mediates the fusion of the viral membrane with specific cellular membranes, resulting in the delivery of the viral genome into the cytosol in the form of free nucleocapsid (6, 17). Enveloped virions belonging to the *Orthomyxoviridae*, *Rhabdoviridae*, and *Togaviridae* enter the cell via coated pits which, upon internalization as vesicles, become part of the endosomal compartment whose contents become acidified (pH ~5) (17). This acidification alters the conformation of the fusion-specific viral glycoprotein, activating its ability to fuse the viral lipid bilayer with that of the endosomal vesicle membrane (4, 8, 10, 15, 16). The viral genome is then liberated into the cytosol.

The specific viral proteins that mediate these fusion events include the hemagglutinin (HA) proteins of type A, B, and C influenza viruses (*Orthomyxoviridae*) and the G glycoprotein of vesicular stomatitis virus (*Rhabdoviridae*) (17). Although the fusion capabilities of HA proteins of influenza viruses are activated at acidic pH, the HA protein, in addition, needs to be proteolytically cleaved into an amino-terminal HA1 and a carboxyl-terminal HA2 to be fusogenic (6, 7, 17).

Metric algorithms can be used to identify all patterns shared by two sequences that satisfy specific local and global criteria of similarity (12). The Sellers TT algorithm provides a description of all patterns shared by two sequences that satisfy the specified length and mismatch density parameters (12). When a metric algorithm is used to detect alignments common between segments of two given sequences, a scoring method must be employed which assesses the likelihood of finding the observed alignment by chance. We have described a scoring function for nucleotide sequences which is the entropy of a particular alignment (9). This scoring method considers a run of identities of length m significantly more important than m runs each of length one identity.

FIG. 1. Comparisons among the amino acid sequences of the HA proteins of type A, B, and C influenza viruses and of PCTP. The amino acid sequences of the three HA proteins (A/Aichi/2/68/X31, B/Lee/40, C/Cal/78) were compared pairwise by using the TT algorithm at settings of mismatch density $r = 0.9$, adjusted length $s = 5$, and gap penalty $n = 1$. The pairwise metric alignments with the best pattern scores were then manually aligned to maximize the similarity among all three sequences. A section of this alignment is illustrated. Similarly, the sequence of PCTP was compared with that of HA (type A), and the metric alignment with the best pattern score is illustrated. Asterisks indicate aligned amino acids identical among three or more sequences; boxes with broken lines draw attention to conserved Cys residues; and boxes with solid lines in the PCTP region illustrate residues common to all four aligned sequences. The residues are numbered as in references 5, 11, and 13.

This scoring method for nucleic acid sequences can be modified in a simple fashion to make it applicable to amino acid sequence analyses. Since there are 20 amino acids, the scoring function takes the form

$$\Pi = \log_{20} \left[\sum_{i=2} 20^{m_i} \right]$$

where m_i is the length of the ith longest run of identities in a given sequence alignment. The summation excludes the length of the longest run in an alignment (the index run). This strategy allows one to focus attention on the alignment path with the highest entropy (pattern) score in each output. An important aspect of the Π function is that it allows for a statistical interpretation of a particular alignment.

The Sellers TT algorithm, suitably modified to score all amino acid matches equally, was used to obtain the alignment of the hemagglutinins of influenza viruses A, B, and C (5, 11, 13). There appears to be a segmental duplication at the amino-terminal end of the HA of influenza C virus. Nine Cys residues are common to all three sequences. These conserved Cys residues include the two that hold HA1 and HA2 together. Approximately 17% of aligned amino acid residues are constant and conserved among all three sequences in the HA2 region. These constant amino acids include Ile, Leu, and Glu (Fig. 1).

The G protein of vesicular stomatitis virus, like the hemagglutinin of influenza viruses, exhibits membrane fusion activity at acidic pH. These sequences (3, 5, 11, 13) share approximately 28% aligned identical residues (between HA A and G). It is noteworthy that a clustering of conserved acidic residues is observed in the region corresponding to the middle of HA2 (Fig. 2). A similar conservation of acidic amino acids is also observed when the G protein of VSV is compared with the HA of type B or of type C influenza viruses.

The amino acid sequence of HA (type A) was compared with that of a water-soluble phosphatidylcholine-specific transport protein (PCTP) and a nonspecific lipid transport protein, both isolated from bovine liver (1, 14). The

```
HA A 410 QIEKEFSEVE GRIQDLEKYVEDTKIDLWSYNAELLVALENQHTIDLTDSEMNKLFEKTR RQLRE NAEEMGNGCFKIYHKCD
         *   * **** *    * *    **      *   *        *   * ** *  *  *     *   *     *    * *
VSV G 378 QWFP FGEVEIGPNGVL K    TKQG   YKFPLHIIG    TGEV DSDI KM E  RVVKHWEHPHIE AAQTF  LKKD
              # #                                   #       #            #
```

```
491 NACIESIRNGTYDHDVYRD EAL NNRFQIKGVELKSGYKDWILWISFAISCFL  LCVVL   LGFIMWACQRGNIRCNICI     566
     *     *         *   * *  *  * * ** **  **  *  * ** * *        * *
441 DTG EVLYYG DTGVSKNPVE LVEGWFSGWRSSLM GVLAVI  IGFVILMFLIKLIGVLSSL FRPK   RRPIYKSDVEMAHFR 517
```

FIG. 2. Comparison between the amino acid sequences of HA of influenza virus (type A) and the G protein of vesicular stomatitis virus (New Jersey). The two sequences were compared as indicated in the legend to Fig. 1. A section of the alignment with the best pattern score is illustrated. Asterisks indicate aligned identical residues; acidic residues that are aligned are boxed in; the symbol # indicates acidic residues that are also aligned when G is compared with either the HA B or the HA C sequences.

alignment between HA and a section of PCTP is also shown in Fig. 1. The segment of PCTP aligned with HA represents the fatty acid-binding site in PCTP (14). The relative positions of several of the acidic residues appear to be similar. This is particularly noteworthy in that PCTP has a pH optimum of approximately 5 (1, 14), as does the fusion activity of the viral proteins (6, 17). The location of the "PCTP-like" region in the three-dimensional structure of the influenza A virus hemagglutinin is the region that extends down the coiled-coil α helices of the trimer stalk (18; Fig. 3). Interspersed down these core helices are the acidic residues (Glu and Asp) that are conserved among the acidic pH-dependent viral fusion-mediating proteins and PCTP. Tyr-54 in PCTP can be cross-linked to a fatty acid in phosphatidylcholine, thus identifying this region as a fatty acid-binding site (14). It is noteworthy that the acidic side chains of Glu and Asp residues embedded in a protein have a pK_a of approximately 5. Thus, at acidic pH these residues are more likely to be discharged and to become more lipophilic.

The sequence alignments observed between PCTP and the influenza HA suggest that a lipid transport activity might be associated with the fusion proteins of enveloped virions. It has been demonstrated by use of lipid monolayer methods that there is lipid transport activity associated with the F protein of Sendai virus (R. Demel and F. R. Landsberger, unpublished data). When a lipid monolayer containing radioactive phosphatidylcholine is exposed to fusion-competent Sendai virus, radioactivity is removed from the monolayer into the aqueous subphase below the monolayer. In similar experiments, purified influenza virions have been found to display phosphatidylcholine transport activity only in the acidic pH range.

A series of mutants of influenza virus that have a higher pH optimum for fusion activity have been characterized (2). Many of these mutants result from the replacement of acidic residues in the "PCTP-like" region. For example, the replacement of Glu-81, Asp-112, or Glu-114 (numbered in HA2) with other amino acids increases the optimum pH at which these mutant virions fuse with cellular membranes. Other mutants involve alterations in amino acids that interact with acidic residues in the PCTP-like region of HA2.

These correlations suggest that the fusion process may be described as (i) binding of the virus to the target membrane, thus bringing the two lipid

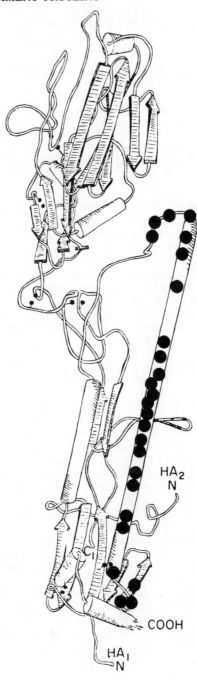

FIG. 3. Schematic drawing showing the three-dimensional structure of a representative type A HA molecule (adapted from reference 18). The structure of a monomer of HA is illustrated. The carboxyl terminus of HA2 is inserted into the viral lipid bilayer at the bottom of the illustration. Flat twisted arrows represent extended β chains, cylinders represent helices, and small filled circles with a tail represnt disulfide bonds. Large filled circles represent amino acids that are aligned and identical between HA (type A) and PCTP (see Fig. 1). Aligned acidic residues are also indicated by large filled circles.

---

bilayers into close apposition; (ii) a pH-determined "discharging" of the acidic amino acids in the PCTP-like lipid binding region; and (iii) destabilization of the target membrane by withdrawing a lipid from it and bringing it over to the viral lipid bilayer. The intermixing of the phospholipids of the two bilayers may be described as the result of the destabilization of the target membrane by withdrawal of lipid in a process resembling that of the molecular events associated with PCTP-mediated lipid transport.

We thank Igor Tamm, Purnell W. Choppin, and Peter H. Sellers for numerous helpful and stimulating discussions.

F.R.L. is an Andrew W. Mellon Fellow and is supported by Public Health Service research grant GM 31790 from the National Institute of General Medical Sciences and grant PCM 8409213 from the National Science Foundation. P.B.S. was supported by Public Health Service research grant AI-16262 from the National Institute of Allergy and Infectious Diseases, a contract from the National Foundation for Cancer Reserarch, an Irma T. Hirschl Award, and an Established Investigatorship from the American Heart Association.

LITERATURE CITED

1. **Akeroyd, R., P. Moonen, J. Westerman, W. C. Puyk, and K. W. A. Wirtz.** 1983. The complete primary structure of the phosphatidylcholine-transfer protein from bovine liver. Eur. J. Biochem. **114**:385–391.
2. **Daniels, R. S., J. C. Downie, A. J. Hay, M. Knossow, J. J. Skehel, M. L. Wang, and D. C. Wiley.** 1985. Fusion mutants of the influenza virus hemagglutinin glycoprotein. Cell **40**:431–439.
3. **Gallione, C. J., and J. K. Rose.** 1983. Nucleotide sequence of a cDNA clone encoding the entire glycoprotein from the New Jersey serotype of vesicular stomatitis virus. J. Virol. **46**:162–169.
4. **Helenius, A., J. Kartenbeck, K. Simons, and E. Fries.** 1980. On the entry of Semliki Forest virus into BHK-21 cells. J. Cell Biol. **84**:404–420.
5. **Krystal, M., R. M. Elliott, E. W. Benz, Jr., J. F. Young, and P. Palese.** 1982. Evolution of influenza A and B viruses: conservation of structural features in the hemagglutinin genes. Proc. Natl. Acad. Sci. USA **79**:4800–4804.
6. **Lamb, R. A., and P. W. Choppin.** 1983. The gene structure and replication of influenza virus. Annu. Rev. Biochem. **52**:467–506.
7. **Lazarowitz, S. G., and P. W. Choppin.** 1975. Enhancement of the infectivity of influenza A and B viruses by proteolytic cleavage of the hemagglutinin polypeptide. Virology **68**:440–454.
8. **Matlin, K. S., H. Reggio, A. Helenius, and K. Simons.** 1981. Infectious entry pathway of influenza virus in a canine kidney cell line. J. Cell Biol. **91**:601–613.
9. **May, L. T., F. R. Landsberger, M. Inouye, and P. B. Sehgal.** 1985. Significance of similarities in patterns: an application to β interferon-related DNA on human chromosome 2. Proc. Natl. Acad. Sci. USA **82**:4090–4094.
10. **Miller, D. K., and J. Lenard.** 1980. Inhibition of vesicular stomatitis virus infection by spike glycoprotein. J. Cell Biol. **84**:430–437.
11. **Nakada, S., R. S. Creager, M. Krystal, R. P. Aaronson, and P. Palese.** 1984. Influenza C virus hemagglutinin: comparison with influenza A and B virus hemagglutinins. J. Virol. **50**:118–124.
12. **Sellers, P. H.** 1984. Pattern recognition in genetic sequences by mismatch density. Bull. Math. Biol. **46**:501–514.

13. **Verhoeyen, M., R. Fang, W. M. Jou, R. Devos, D. Huylebroeck, E. Saman, and W. Fiers.** 1980. Antigenic drift between the hemagglutinin of the Hong Kong influenza strains A/Aichi/2/68 and A/Victoria/3/75. Nature (London) **286:**771–776.

14. **Westerman, J., K. W. A. Wirtz, T. Berkhout, L. L. M. van Deenen, R. Radhakrishnan, and H. G. Khorana.** 1983. Identification of the lipid binding site of phosphatidylcholine-transfer protein with phosphatidylcholine analogs containing photoactivable carbene precursors. Eur. J. Biochem. **132:**441–449.

15. **White, J., and A. Helenius.** 1980. pH-dependent fusion between the Semliki Forest virus membrane and liposomes. Proc. Natl. Acad. Sci. USA **77:**3273–3277.

16. **White, J., J. Kartenbeck, and A. Helenius.** 1980. Fusion of Semliki Forest virus with plasma membrane can be induced by low pH. J. Cell Biol. **87:**264–272.

17. **Wiley, D. C.** 1985. Viral membranes, p. 45–67. *In* B. N. Fields (ed.), Virology. Raven Press, New York.

18. **Wilson, I. A., J. J. Skehel, and D. C. Wiley.** 1981. Structure of the haemagglutinin membrane glycoprotein of influenza virus at 3 A resolution. Nature (London) **289:**366–373.

Role of the Viral Envelope in Leukemia Caused by Friend Murine Leukemia Virus

ALLEN OLIFF†

DeWitt Wallace Research Laboratory, Memorial Sloan Kettering Cancer Center, New York, New York 10021

Friend murine leukemia virus (F-MuLV) is a replication-competent retrovirus that induces erythroleukemia in susceptible strains of mice. F-MuLV-induced erythroleukemias are invariably associated with the generation of a second retrovirus, Friend mink cell focus-inducing virus (Fr-MCF). Fr-MCF is generated by recombination between F-MuLV and endogenous MCF-like sequences present in normal mouse DNA. Mice that are resistant to F-MuLV-induced leukemia permit replication of F-MuLV but do not form Fr-MCF. Clonal isolates of Fr-MCF also cause erythroleukemia, but no F-MuLV-like viruses are found in Fr-MCF-induced leukemias. These results suggest that Fr-MCF is responsible for F-MuLV-induced leukemia. Molecular analysis of F-MuLV and Fr-MCF DNAs indicate that the major difference between these viruses resides in the envelope gene. These sequences control the host-range properties of F-MuLV and Fr-MCF. To define the viral sequences that determine leukemogenicity, I constructed hybrid viral DNAs between F-MuLV or Fr-MCF and a nonleukemogenic retrovirus, amphotropic virus (Ampho). These hybrid DNAs produce recombinant viruses upon transfection into fibroblasts. F-MuLV/Ampho and Fr-MCF/Ampho recombinants require the F-MuLV or Fr-MCF envelope genes to cause leukemia. However, the F-MuLV/Ampho recombinants that cause leukemia always generate MCF viruses in diseased mice. We conclude that the Fr-MCF envelope gene contributes to both F-MuLV- and Fr-MCF-induced leukemogenesis.

Murine erythroleukemia-inducing viruses were first isolated in 1957 by Friend, who found that cell-free filtrates from an Ehrlich ascites tumor produced a fatal erythroleukemia in Swiss mice (10). Repeated passages of cell-free filtrates from the leukemic tissue of these animals into uninfected mice produced a stable preparation of highly leukemogenic virus, Friend virus. Subsequent studies by other investigators found that all Friend virus isolates actually contained two viral components: a replication-competent type C retrovirus, Friend murine leukemia virus (F-MuLV), and a replication-defective retrovirus, spleen focus-forming virus (SFFV) (26, 34, 38). Together these viruses (F-MuLV and SFFV) are designated the Friend virus complex. Toxler and Scolnick succeeded in isolating the individual components of the Friend virus complex in clonal fibroblast cell cultures in 1977 (35, 36). Once isolated, the individual

†Present address: Merck Sharp & Dohme Research Laboratories, West Point, PA 19486.

viruses were tested for the ability to cause leukemia independently of one another. SFFV was found to be responsible for the rapid (1 to 2 weeks) erythroleukemia seen in adult mice after inoculation with the Friend virus complex. F-MuLV is also highly leukemogenic, but F-MuLV causes disease only if inoculated into newborn mice. F-MuLV disease develops in 6 to 8 weeks after inoculation and affects virtually 100% of infected animals (20, 36). This rapid F-MuLV disease manifests as an erythroleukemia. However, F-MuLV can also induce myeloid and lymphoid leukemias with a longer latency period (5, 32).

The genomes of both SFFV and F-MuLV were molecularly cloned in 1978 (19, 21). The application of recombinant DNA techniques to retroviral genetics makes it possible to analyze the Friend virus genome to identify the viral genes responsible for leukemia. Molecularly cloned F-MuLV DNA contains the same viral genes as other replication-competent murine retroviruses, long terminal repeat (LTR) *gag pol env* LTR (15, 16, 33). No oncogenes are present in the F-MuLV genome, and no transduced cellular sequences or unusual viral sequences have been found. Why, then, is F-MuLV highly pathogenic? What viral genes contribute to the leukemogenic phenotype of F-MuLV?

To answer this question, my colleagues and I began a series of experiments in 1980 aimed at identifying the smallest possible segment of F-MuLV DNA that would generate a leukemogenic virus in conjunction with an otherwise nonpathogenic murine leukemia virus (MuLV). We were aided in this analysis by the isolation of a wild mouse amphotropic retrovirus (Ampho) that is nonleukemogenic in NIH Swiss mice (4, 27). Ampho is also a replication-competent type C retrovirus that replicates to high titers (10^4 infectious units/ml) in vitro and in vivo. More importantly, Ampho replicates in bone marrow cultures and can infect murine erythroid precursors (38). However, newborn NIH Swiss or NFS/n mice inoculated with Ampho do not develop leukemia or any other hematopoietic proliferative disorders (22, 23). Therefore, leukemogenic recombinant viruses formed between Ampho and pieces of F-MuLV DNA must acquire their pathogenic characteristics from the F-MuLV sequences used to create these new viruses.

One simple yet precise method for generating F-MuLV/Ampho recombinants is to use common restriction enzyme sites to exchange homologous segments of DNA between the genomes of these viruses. Figure 1 depicts the restriction enzyme maps of molecularly cloned F-MuLV, Ampho, and Friend mink cell focus-inducing virus (Fr-MCF) DNAs. The viral DNAs are presented in permuted orientations relative to one another. Nonetheless, several enzyme sites are clearly present at the same positions in all three DNAs (e.g., *Cla*I and *Sph*I). Using these sites, we constructed a series of hybrid viral DNAs that contain specific sequences from each virus (Fig. 2). An example of the cloning strategy used to construct one of these hybrid DNAs is depicted in Fig. 3. The F-MuLV 5' LTR and the *gag* and *pol* genes located between the *Cla*I and *Sph*I sites are joined to the Ampho envelope gene and 3' LTR between the *Sph*I and *Pst*I sites. Further digestion with *Cla*I and subcloning into pBR322 yields the F/A ENV hybrid DNA that contains the F-MuLV LTR and *gag* and *pol* genes and the Ampho envelope gene.

Once isolated, these hybrid DNAs are transfected into NIH 3T3 cells where they direct the synthesis of recombinant retroviruses. Retroviruses isolated in

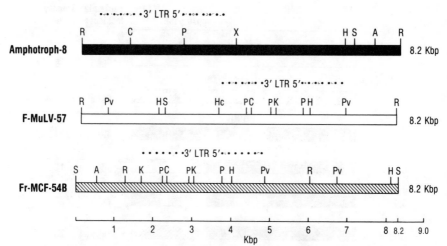

FIG. 1. Schematic representation of the restriction endonuclease maps of molecularly cloned F-MuLV, Fr-MCF, and Ampho DNAs. Abbreviations: A, *Acc*I; C, *Cla*I; H, *Hind*III; Hc, *Hinc*II; K, *Kpn*I; P, *Pst*I; Pv, *Pvu*I; R, *Eco*RI; S, *Sph*I; X, *Xba*I; Kbp, kilobase pairs.

these experiments exhibit the growth characteristics expected of recombinant viruses derived from specific segments of F-MuLV, Fr-MCF, and Ampho DNAs. For example, F-MuLV is an ecotropic (mouse tropic) virus and will only grow on murine cells. Ecotropism is an envelope gene-encoded property (28). The two recombinant viruses that contained the F-MuLV envelope gene (CL-1 and 5A25) also exhibited ecotropism (Table 1). Recombinants that contained the Ampho envelope gene (L1, A/Fr LTR, and F/A ENV) exhibited amphotropism (growth on both murine and nonmurine cells). Newly isolated viruses were tested for the ability to cause leukemia in newborn NFS/n mice. Two to four litters of newborn mice (<24 h old) were inoculated intraperitoneally with 5×10^4 to 1×10^5 infectious units of virus. Representative animals were periodically assessed for signs of hematopoietic pathology by microscopic examination of peripheral blood smears. Grossly diseased animals were killed and subjected to a complete autopsy. The results of these experiments are presented in Fig. 2.

Three points are apparent from the data. (i) Only those viruses that contain the F-MuLV or Fr-MCF envelope genes are leukemogenic (CL-1, 5a25, A/Fr E+L, and L9). However, the Fr-MCF envelope gene by itself is not sufficient to turn Ampho into a pathogenic virus (A/Fr ENV). Similarly, the 3′ LTRs of F-MuLV and Fr-MCF by themselves do not produce leukemogenic recombinants with Ampho (L1 and A/Fr LTR). (ii) The 3′ LTRs of F-MuLV and Fr-MCF can boost the incidence of leukemia exhibited by viruses containing these sequences in addition to the F-MuLV or Fr-MCF envelope genes (A/Fr E+L 14% leukemia versus A/Fr ENV 0% leukemia, and CL-1 50% leukemia versus 5a25 22% leukemia). (iii) Even the most pathogenic recombinants constructed between F-MuLV and Ampho or between Fr-MCF and Ampho do not cause leukemia as

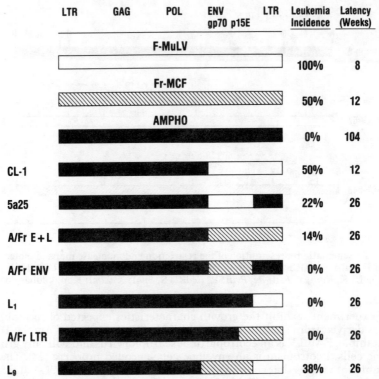

FIG. 2. Schematic representation of recombinant viral genomes constructed between F-MuLV, Fr-MCF, and Ampho DNAs (from reference 23). Maps refer to proviral DNA structure after reverse transcription and integration of the viral DNA into host cells. Leukemia incidence refers to NIH Swiss mice injected as newborns. Latency refers to the maximum time until disease. Many animals develop leukemia before the maximum latent period. No animals injected with Fr-MCF/Ampho or CL-1 developed leukemia between 12 and 26 weeks postinoculation.

frequently or as quickly as wild-type F-MuLV or Fr-MCF. Apparently, other determinants of leukemogenicity must exist outside the envelope gene and LTR regions. To test this hypothesis, we constructed three additional hybrid DNAs and generated three new recombinants (Fig. 4).

Each of these hybrid DNAs contains the F-MuLV 5' LTR and the *gag* and *pol* genes joined to the envelope gene or 3' LTR of a nonpathogenic (Ampho and L1) or weakly pathogenic (L9) virus. All three of the recombinant viruses made from these DNAs cause leukemia. Ampho and L1 are converted into weakly pathogenic viruses, F/A E+L and F/A ENV. L9 is converted into a strongly pathogenic virus, F/Fr ENV. Clearly, the F-MuLV 5' LTR and *gag* and *pol* genes contain determinants of viral pathogenicity that contribute to the leukemogenic phenotype of F-MuLV. These results are similar to those reported by Holland et al.

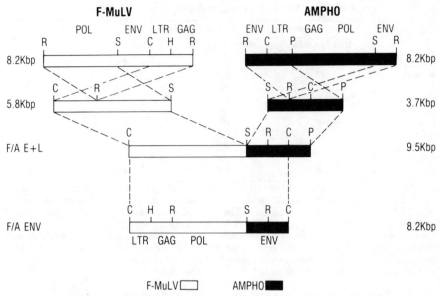

FIG. 3. Schematic representation of the molecular cloning of the recombinant viral genomes F/A E+L and F/A ENV. Each subgenomic viral DNA fragment was cloned into a plasmid vector and analyzed by restriction endonuclease mapping before being used to construct F/A E+L. Abbreviations: C, *Cla*I; H, *Hin*dIII; P, *Pst*I; R, *Eco*RI; S, *Sph*I; Kbp, Kilobase pairs.

(14) concerning the viral sequences responsible for thymic leukemias caused by AKR-MCF 247. In making recombinant viruses between AKR-MCF 247 and AKv virus, Holland found that four distinct regions of the AKR-MCF 247 genome contributed to leukemogenicity. Again, the principal determinants of pathogenicity localize to the envelope gene (gp70 and p15E) and the 3' LTR. But the AKR-MCF 247 *gag* and *pol* sequences also increase the leukemogenicity of recombinant viruses. These four regions appear to be additive in their pathogenic potential. As more of these regions are added to hybrid DNA constructs, the resulting recombinant viruses exhibit greater leukemogenicity. Other investigators have focused on one or another of these regions as the "primary determinant" of retroviral leukemogenesis (6, 7, 18, 29). However, virtually all of these studies find that at least two and as many as four distinct segments of the viral genome contribute to the pathogenic phenotype of murine leukemia viruses.

Rather than asking, "What viral gene is responsible for causing leukemia?" a more appropriate question is, "How do individual viral genes contribute to leukemogenesis?" In the case of F-MuLV we are beginning to dissect this process. The F-MuLV and Fr-MCF LTRs increase the incidence of leukemia and decrease the latency period exhibited by viruses containing these sequences.

TABLE 1. Virus host range and growth characteristics

Virus	Reverse transcriptase activity (IU/ml)[a] in:			XC fusions
	NIH 3T3 cells	BALB/3T3 cells	Mink lung cells	
F-MuLV	10^5	10^5	<1	+
Fr-MCF	10^4	10^4	10^3	−
Ampho	10^4	10^1	10^4	−
CL-1	10^5	10^2	<1	+
5a25	10^5	10^2	<1	+
L-1	10^4	10^2	10^4	−
A/Fr LTR	10^4	10^2	10^4	−
A/Fr Env	10^4	10^1	10^3	−
A/Fr E+L	10^4	10^2	10^3	−
L9	10^4	10^2	10^3	−

[a] The host range of each viral isolate was ascertained by a reverse transcriptase assay. Viral stocks were diluted in serial log dilutions onto 60-mm culture dishes containing 10^5 NIH 3T3, BALB/3T3, or mink lung cells. The cells were allowed to grow to confluence and were passed 1:3 into fresh culture dishes. When the second passage reached confluence, the cells were fed with 5 ml of fresh medium and incubated for 18 h. All 5 ml of medium was removed and used to determine reverse transcriptase activity. Plates that were negative for viral reverse transcriptase were passaged twice more and reassayed to ensure that correct endpoint titers were obtained. IU, Infectious units.

Presumably, the F-MuLV and Fr-MCF LTRs promote retroviral transcription more efficiently than the Ampho LTR in hematopoietic target cells. The sequences responsible for this effect most likely lie within the U3 region of the LTR. Both F-MuLV and Fr-MCF contain direct repeat elements homologous to the transcriptional enhancers described in the Moloney sarcoma virus LTR (16, 17). Interestingly, F-MuLV has two copies of these enhancerlike sequences while Fr-MCF has only one copy. Recombinant viruses constructed with the F-MuLV 3' LTR (L9) are more pathogenic than similar constructs made with the Fr-MCF 3' LTR (A/Fr E+L). The additional copy of enhancer sequences present in the F-MuLV LTR may be responsible for the increased pathogenicity of L9 versus A/Fr E+L. The F-MuLV LTR U3 region is also responsible for the erythroid nature of F-MuLV disease (2). These sequences are able to convert lymphoid leukemia viruses into erythroleukemia-inducing viruses. Similarly, the U3 region of retroviruses that cause thymic leukemia dictates the thymotropism of those viruses (8).

   The F-MuLV envelope gene converts Ampho into a weakly pathogenic virus (5a25). One possible explanation for this effect is that the F-MuLV envelope changes the host range of Ampho and allows the recombinant virus to penetrate a critical target cell that Ampho cannot infect. However, Ampho pseudotypes of SFFV cause erythroleukemia, and these pseudotypes are surrounded entirely by Ampho envelope proteins (1, 38). Therefore, Ampho must be able to penetrate some erythroid precursors. A more attractive hypothesis is that the F-MuLV envelope gene participates in the formation of Fr-MCF virus in infected animals. Mink cell focus-inducing viruses arise via recombination between exogenously infecting retroviruses like F-MuLV and endogenous retroviral sequences present

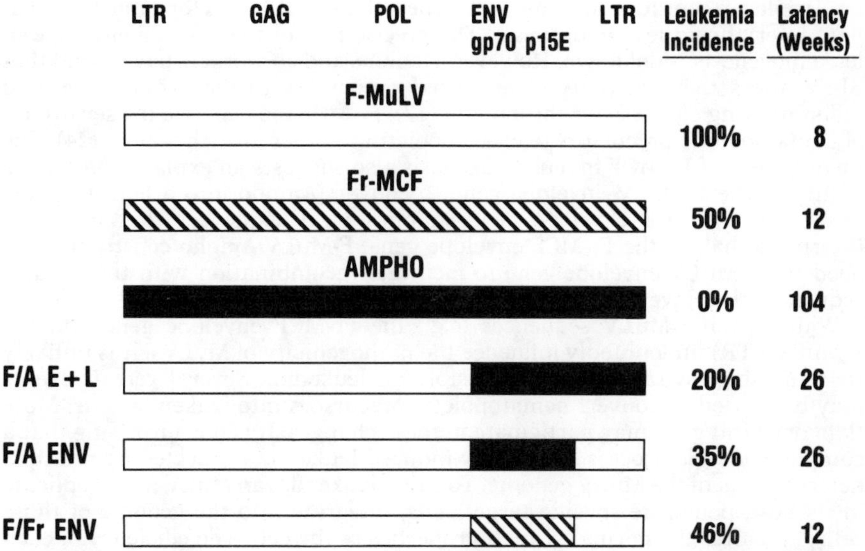

LTR	GAG	POL	ENV gp70 p15E	LTR	Leukemia Incidence	Latency (Weeks)

F-MuLV

100%   8

Fr-MCF

50%   12

AMPHO

0%   104

F/A E+L   20%   26

F/A ENV   35%   26

F/Fr ENV   46%   12

FIG. 4. Schematic representation of the hybrid viral DNAs constructed using the F-MuLV 5.8-kilobase-pair *Cla-Sph* DNA fragment. Maps refer to proviral DNA structure after reverse transcription and integration of the viral DNA into host cells. Leukemogenicity data are taken from Table 1. Latency refers to the maximum time until disease. Many animals developed leukemia before the maximum latency period. No animals injected with Fr-MCF/Ampho or F/Fr ENV/Ampho developed leukemia between 12 and 26 weeks postinoculation.

in the germ line of most strains of mice (3, 9). Fr-MCF is invariably found in the leukemic tissue of F-MuLV-infected mice (37). Fr-MCF is not found in normal hematopoietic tissues, nor is it present in hematopoietic tissues of F-MuLV-infected animals that do not develop leukemia (30). F-MuLV replicates to high titers in adult Swiss mice, but does not generate Fr-MCF in these animals and does not cause leukemia (31). Both biologic and molecular clones of Fr-MCF cause leukemia in adult as well as newborn animals. These observations raise the possibility that Fr-MCF and not F-MuLV is the actual cause of erythroleukemia. Interestingly, all of the pathogenic recombinants constructed with the F-MuLV envelope gene (CL-1 and 5a25) generate mink cell focus-inducing virus when they cause leukemia (22, 23).

Heteroduplex and nucleotide sequence analysis of Fr-MCF and F-MuLV indicate that these viruses are nearly identical throughout their genomes except for the N-terminal half of the envelope gene (11, 16). In this region Fr-MCF diverges from F-MuLV, but is highly homologous to the envelope gene of SFFV. Both SFFV and Fr-MCF induce erythroleukemias in the absence of F-MuLV. F-MuLV does not cause erythroleukemia without SFFV or Fr-MCF. These studies suggest that the "proximal cause" of erythroleukemia in F-MuLV-infected mice is Fr-MCF. Furthermore, since the F-MuLV and Fr-MCF genomes

are similar except for their envelope genes, the Fr-MCF envelope must participate in erythroid leukemogenesis. The precise role of the Fr-MCF envelope in
leukemogenesis is unknown. However, Hankins and co-workers have found that
SFFV alters the sensitivity of erythroid precursors to the proliferative and
differentiating effects of erythropoietin (12). Fr-MCF may change the sensitivity
of hematopoietic precursors to erythropoietin or other growth factors (24). The
involvement of Fr-MCF in leukemogenesis also suggests an explanation for the
ability of the F-MuLV envelope gene to convert Ampho into a leukemogenic
virus. The F-MuLV envelope gene supplies viral sequences that form the
C-terminal half of the Fr-MCF envelope gene. F-MuLV/Ampho constructs may
need the F-MuLV envelope gene to facilitate recombination with the cellular
sequences that give rise to Fr-MCF.

While specific MuLV sequences (e.g., the Fr-MCF envelope gene and the
F-MuLV LTR) undoubtedly influence the pathogenicity of MuLVs, it is unlikely
that any single viral gene is responsible for leukemia. Several genetic events
may be needed to convert hematopoietic precursors into leukemia cells. More
than one viral gene may participate in these changes. It is not surprising that a
complex biologic process like virally induced leukemia is affected by multiple
determinants in the MuLV genome. To cause leukemia, an MuLV must replicate
in its host, penetrate specific target cells, integrate into the genome of those
cells, and alter the normal growth properties of the cell. The efficiency of each
of these functions will influence the overall ability of an MuLV to cause disease.
Clonal isolates of F-MuLV and other highly leukemogenic MuLVs (e.g., AKR-
MCF 247 and SL3-3) have been selected for their rapid and efficient induction of
leukemias (13, 25). These pathogenic isolates may have maximized their ability
to cause leukemia by acquiring advantageous mutations throughout their
genomes.

We thank Paul Marks and Richard Rifkind for their advice and support in carrying out these
experiments.
These studies were supported in part by the David Schwartz Foundation and the Kleberg
Foundation, by grant MV-170 from the American Cancer Society, and by Public Health Service grant
CA 08748 from the National Cancer Institute.

LITERATURE CITED

1. **Bestwick, R., M. Ruta, A. Kiessling, C. Faust, D. Linemeyer, E. Scolnick, and D. Kabat.** 1983.
   Genetic structure of Rauscher spleen focus-forming virus. J. Virol. **45**:1217–1222.
2. **Chatis, P. A., C. A. Holland, J. W. Hartley, and W. P. Rowe.** 1983. Role for the 3' end of the genome
   in determining disease specificity of Friend and Moloney murine leukemia viruses. Proc. Natl.
   Acad. Sci. USA **80**:4408–4411.
3. **Chattopadhyay, S. K., M. R. Lander, E. Rands, and D. R. Lowry.** 1980. Structure of endogenous
   murine leukemia virus DNA in mouse genomes. Proc. Natl. Acad. Sci. USA **77**:5774–5778.
4. **Chattopadhyay, S. K., A. Oliff, D. Linemeyer, M. R. Lander, and D. R. Lowy.** 1981. Genomes of
   murine leukemia viruses isolated from wild mice. J. Virol. **39**:777–791.
5. **Chesebro, B., J. L. Portis, K. Wehrly, and J. Nishio.** 1983. Effect of murine hot genotype on MCF
   virus expression, latency, and leukemia cell type of leukemias induced by Friend leukemia helper
   virus. Virology **128**:221–233.
6. **DesGroseillers, L., and P. Jolicoeur.** 1984. Mapping the viral sequences conferring leukemogenicity
   and disease specificity in Maloney and amphotropic murine leukemia viruses. J. Virol. **52**:448–456.
7. **DesGroseillers, L., and P. Jolicoeur.** 1984. The tandem direct repeats within the long terminal
   repeat of murine leukemia viruses are the primary determinant of their leukemogenic potential. J.
   Virol. **52**:945–952.

8. **DesGrosiellers, L., E. Rassart, and P. Jolicoeur.** 1983. Thymotropism of murine leukemia virus is conferred by its long terminal repeat. Proc. Natl. Acad. Sci. USA **80**:4203–4207.

9. **Elder, J. H, J. W. Gautsch, F. C. Jensen, R. A. Lerner, J. W. Hartley, and W. P. Rowe.** 1977. Biochemical evidence that MCF murine leukemia viruses are envelope (env) gene recombinants. Proc. Natl. Acad. Sci. USA **74**:4676–4680.

10. **Friend, C.** 1957. Cell-free transmission in adult Swiss mice of a disease having the character of a leukemia. J. Exp. Med. **105**:307–318.

11. **Gonda, M. A., J. Kaminchick, A. Oliff, J. Menke, K. Nagashima, and E. M. Scolnick.** 1984. Heteroduplex analysis of molecular clones of the pathogenic Friend virus complex: Friend murine leukemia, mink cell focus-forming, and the polycythemia- and anemia-inducing strains of spleen focus-forming viruses. J. Virol. **51**:306–314.

12. **Hankins, W. D.** 1983. Increased erythropoietin sensitivity after in vitro transformation of hematopoietic precursors by RNA tumor viruses. J. Natl. Cancer Inst. **70**:725–729.

13. **Hartley, J. W., N. K. Wolford, L. J. Old, and W. P. Rowe.** 1977. A new class of murine leukemia virus associated with development of spontaneous lymphomas. Proc. Natl. Acad. Sci. USA **74**:789–792.

14. **Holland, C. A., J. W. Hartley, W. P. Rowe, and N. Hopkins.** 1985. At least four viral genes contribute to the leukemogenicity of murine retrovirus MCF 247 in AKR mice. J. Virol. **53**:158–165.

15. **Koch, W., G. Hunsmann, and R. Friedrich.** 1983. Nucleotide sequence of the envelope gene of Friend murine leukemia virus. J. Virol. **45**:1–9.

16. **Koch, W., W. Zimmerman, A. Oliff, and R. Friedrich.** 1984. Molecular analysis of the envelope gene and long terminal repeat of Friend mink cell focus-inducing virus: implications for the functions of these sequences. J. Virol. **49**:828–840.

17. **Laimins, L. A., P. Gruss, R. Pozzatti, and G. Khoury.** 1984. Characterization of enhancer elements in the long terminal repeat of Moloney murine sarcoma virus. J. Virol. **49**:183–189.

18. **Lentz, J., D. Celander, R. L. Crowther, R. Patarca, and A. Haseltine.** 1984. Determination of the leukaemogenicity of a murine retrovirus by sequences within the long terminal repeat. Nature (London) **308**:467–469.

19. **Linemeyer, D. L., S. K. Ruscetti, J. G. Menke, and E. M. Scolnick.** 1980. Recovery of biologically active spleen focus-forming virus from molecularly cloned spleen focus-forming virus-pBR322 circular DNA by cotransfection with infectious type C retroviral DNA. J. Virol. **35**:710–721.

20. **MacDonald, M. R., T. W. Mak, and A. Bernstein.** 1980. Erythroleukemia induction by replication competent type-C viruses cloned from the anemia and polycythemia inducing isolates of Friend leukemia virus. J. Exp. Med. **151**:1493–1503.

21. **Oliff, A. I., G. L. Hager, E. H. Chang, E. M. Scolnick, H. W. Chan, and D. R. Lowy.** 1980. Transfection of molecularly cloned Friend murine leukemia virus DNA yields a highly leukemogenic helper-independent type C virus. J. Virol. **33**:475–486.

22. **Oliff, A., and S. Ruscetti.** 1983. A 2.4-kilobase-pair fragment of the Friend murine leukemia virus genome contains the sequences responsible for Friend murine leukemia virus-induced erythroleukemia. J. Virol. **46**:718–725.

23. **Oliff, A., K. Signorelli, and L. Collins.** 1984. The envelope gene and long terminal repeat sequences contribute to the pathogenic phenotype of helper-independent Friend viruses. J. Virol. **51**:788–794.

24. **Oliff, A., I. Oliff, B. Schmidt, and N. Famulari.** 1984. Isolation of hematopoietic cell lines from the first stage of F-MuLV induced leukemia. Proc. Natl. Acad. Sci. USA **81**:5464–5467.

25. **Pedersen, F. S., R. Crowther, D. Tenney, A. M. Reimold, and W. A. Haseltine.** 1981. Novel leukemogenic retroviruses isolated from a cell line derived from spontaneous AKR tumor. Nature (London) **292**:167–170.

26. **Pluznik, D. H., and L. Sachs.** 1964. Quantitation of a murine leukemia virus with a spleen colony assay. J. Natl. Cancer Inst. **35**:535–564.

27. **Rasheed, S., M. B. Gardner, and E. Chan.** 1976. Amphotropic host range of naturally occurring wild mouse leukemia viruses. J. Virol. **19**:13–18.

28. **Rein, A.** 1982. Interference grouping of murine leukemia viruses: a distinct receptor for the MCF-recombinant viruses in mouse cells. Virology **120**:251–257.

29. **Robinson, H. L., B. M. Blais, P. M. Tsichlis, and J. M. Coffin.** 1982. At least two regions of the viral genome determine the oncogenic potential of avian leukosis viruses. Proc. Natl. Acad. Sci. USA **79**:1225–1229.

30. **Ruscetti, S., L. Davis, J. Fields, and A. Oliff.** 1981. Friend MuLV-induced leukemia is mink cell focus-inducing (MCF) virus associated and is genetically restricted in mice expressing endogenous xeno-related envelope viral genes. J. Exp. Med. **154**:907–920.

31. **Ruscetti, S., J. Fields, L. Davis, and A. Oliff.** 1982. Factors determining the susceptibility of NIH Swiss mice to erythroleukemia induced by Friend murine leukemia virus. Virology **117:**357–365.
32. **Shibuya, T., and T. W. Mak.** 1982. Host control of susceptibility to erythroleukemia and to the types of leukemia induced by Friend murine leukemia virus: initial and late stages. Cell **31:**483–493.
33. **Shinnick, T. M., R. A. Lerner, and J. G. Sutcliffe.** 1981. Nucleotide sequence of Moloney murine leukaemia virus. Nature (London) **293:**543–548.
34. **Steeves, R. A.** 1975. Spleen focus-forming virus in Friend and Rauscher leukemia virus preparations. J. Natl. Cancer. Inst. **54:**289–297.
35. **Toxler, D. H., W. P. Parks, W. C. Vass, and E. M. Scolnick.** 1977. Isolation of a fibroblast nonproducer cell line containing the Friend strain of the spleen focus-forming virus. Virology **76:**602–615.
36. **Troxler, D. H., and E. M. Scolnick.** 1978. Rapid leukemia induced by cloned Friend strain of replication competent murine type-C virus. Virology **85:**17–27.
37. **Troxler, D. H., E. Yuan, D. Linemeyer, S. Ruscetti, and E. M. Scolnick.** 1978. Helper-independent mink cell focus-inducing strains of Friend murine type-C virus: potential relationship to the origin of replication-defective spleen focus-forming virus. J. Exp. Med. **148:**639–653.
38. **Troxler, D. H., S. Ruscetti, D. L. Linemeyer, and E. M. Scolnick.** 1980. Helper-independent and replication-defective erythroblastosis-inducing viruses contained within anemia-inducing Friend virus complex. Virology **102:**28–45.

Cellular Receptors

Purification of a Membrane Receptor Protein for the Group B Coxsackieviruses

JOHN E. MAPOLES,† DAVID L. KRAH,‡ AND RICHARD L. CROWELL

Department of Microbiology and Immunology, Hahnemann University School of Medicine, Philadelphia, Pennsylvania 19102

We have extracted the virus-receptor complex (VRC) formed between the picornavirus coxsackievirus B3 (CB3) and its specific cell receptor on HeLa cells, using the detergents sodium deoxycholate and Triton X-100. The VRC was identified by its sedimentation coefficient (140S), which was less than that of virions (155S). Formation of the VRC from cell lysates and ^{3}H-CB3 occurred with the same specificity as attachment of virions to cells, in that its formation was blocked by unlabeled CB3, but not by poliovirus. The VRC was purified 30,000-fold by differential and sucrose gradient centrifugation. Subsequent iodination with Na^{125}I revealed that the VRC consisted of the normal CB3 proteins and only one additional protein (Rp-a), which had a molecular weight of 49,500. This latter protein was eluted from the VRC and was found to rebind to CB3 and CB1 virions, but not to poliovirus T1. Rp-a is considered to be the protein in the membrane receptor complex responsible for the specific recognition and binding of the group B coxsackieviruses.

The group B coxsackieviruses are human enteroviruses of the picornavirus family which attach to receptor sites located on the surface of the plasma membrane of susceptible cells as a prerequisite to penetration, eclipse, and uncoating of the viral genome (2, 9). These receptor sites are specific for the group B coxsackieviruses and do not bind other enteroviruses such as poliovirus, echoviruses, or the group A coxsackieviruses (1). The HeLa cell contains about 10^5 copies of the coxsackievirus B3 (CB3) receptor, which we have proposed to be a complex of intrinsic plasma membrane glycoproteins (6).

Most receptors constitute only a small fraction of the total membrane protein of the cell, and their purification can be difficult. The purification of some membrane receptors has been facilitated by binding detergent-solubilized membrane preparations to specific ligands which have been immobilized on inert surfaces, as in affinity chromatography (5, 11). Virions can be used to prepare affinity columns for isolation of viral receptors, as has been done with adenovirus (4, 12). Svensson et al. (12) cross-linked adenovirus and bound the particles to AH-Sepharose 4B. Using this technique together with chromatography on wheat germ lectin-Sepharose 4B, these workers isolated two polypeptides of mass 40,000 and 42,000 daltons, which they suggested were the

---

† Present address: University of Colorado Health Sciences Center, Denver, CO 80262.
‡ Present address: Rockefeller University, New York, NY 10021.

adenovirus receptor attachment proteins. A difficulty with this approach for coxsackievirus is that virions are frequently degraded by the binding protocols.

However, the virions in solution can be viewed as affinity particles which specifically bind the solubilized receptor protein as a ligand. The affinity particle can be purifed by traditional methods used for virus purification because the physical characteristics of the virus-receptor complex (VRC) are very similar to those of free virions. This approach to receptor purification was based on the work of Lonberg-Holm et al. with poliovirus (8). When the polio VRC was extracted from cells and centrifuged on a sucrose gradient, the virion sedimentation coefficient decreased from 156S to 130–135S as a result of bound cellular material. Attempts to identify the cellular material were unsuccessful because of contamination by other cellular protein complexes in the sucrose gradient. However, the approach was appealing for three reasons. First, the VRC of CB3 is more resistant to detergent dissociation than the polio VRC, allowing use of stronger detergent mixtures to disrupt contaminating protein complexes. Second, the relatively large sedimentation coefficient of the VRC allows use of differential and sucrose gradient sedimentation to separate the particles from the detergent-solubilized cell material. Third, labeled virus ([³H]leucine) is a convenient marker to monitor VRC purification.

PURIFICATION OF THE HeLa CELL RECEPTOR FOR CB3 (10)

Earlier studies from our laboratory showed that 0.2% deoxycholate solubilizes the HeLa cell plasma membrane receptor for binding and eclipsing CB3 (3). In the present study, further experiments demonstrated that 1.0% Triton X-100 plus 0.5% deoxycholate extracted optimum amounts of cell-associated virus from native HeLa cells. Figure 1 demonstrates that the virus extracted from the solubilized cells sedimented more slowly (140S) than the marker virus (155S). This decrease in sedmientation coefficient was reversed by 1.0% sodium dodecyl sulfate, which dissociates the virus-receptor bond when the complex is suspended in phosphate-buffered saline. The specificity of the formation of the 140S complex was tested by incubating unlabeled CB3 with a cell lysate overnight (this also forms VRC and will saturate the receptors). We found that preincubation of the extract with unlabeled CB3 (2×10^5 virions per cell equivalent) completely eliminated the formation of ³H-CB3 VRC (the 140S particle). A similar incubation with unlabeled poliovirus had no effect. We concluded that the 140S particle consisted of a virion specifically bound to its receptor in the form of a VRC.

Because of the difficulty in labeling cell protein in vivo to high specific activity (the receptor constitutes about 0.002% of the cell protein), we identified the bound receptor protein by iodinating purified VRC. Prior to iodination, VRC was purified by twice sedimenting onto metrizamide pads followed by sucrose gradient centrifugation. An important feature of the purification procedure was that 1.0% sodium dodecyl sulfate was added to the cell lysate before purification, to eliminate small amounts of aggregated protein which interfered with subsequent iodination and immunoprecipitation of the VRC. Although 1% sodium dodecyl sulfate disrupted the virus-receptor bond of isolated VRC, the crude cell lysate contained sufficient solubilized lipids, protein, and nonionic

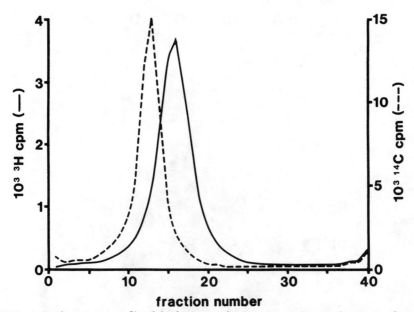

FIG. 1. Radioactivity profile of the fractions of a 5 to 20% sucrose gradient centrifuged with a sample of extracted ³H-cell-associated virus and ¹⁴C-CB3 marker virus. Cell-associated virus was extracted with 1.0% Titron X-100 and 0.5% deoxycholate, the nuclei were removed, and a 10-μl sample was added to 90 μl of buffer and centrifuged with marker virus. Sedimentation was from right to left. Extracted ³H-cell-associated virus (———), ¹⁴C-CB3 marker (——). (From reference 10 with permission.)

detergents to prevent this disruption. The purified VRC had the same specific activity (PFU per counts per minute) as the original virus and sedimented more slowly (145S) than marker virus. The efficiency of the purification procedure was about 3×10^4 (determined from the ratio of the recoveries of the final VRC sample and a control sample of labeled cells). The cellular receptor protein bound to CB3 was then detected by iodination of the VRC with chloramine-T. The sedimentation value of the iodinated VRC remained unchanged, although the infectivity of the carrier virus was partially destroyed in the labeling process.

Analysis by electrophoresis showed that the iodinated VRC contained the major capsid proteins of CB3 as well as one additional protein, with a molecular weight of 49,500 (Fig. 2, lane B). This protein was designated as the receptor protein a (Rp-a). Lane C of Fig. 2 is an iodine-labeled control sample prepared from a cell lysate without added CB3 and purified in parallel with the VRC. Rp-a is absent, and most of the iodinated material appears to be irreversibly aggregated. The labeled VRC could be immunoprecipated with rabbit anti-CB3 serum (Fig. 2, lane D), and it is clear that Rp-a was precipitated with the virus. These results demonstrate that Rp-a was bound to the virus, rather than copurifying with it.

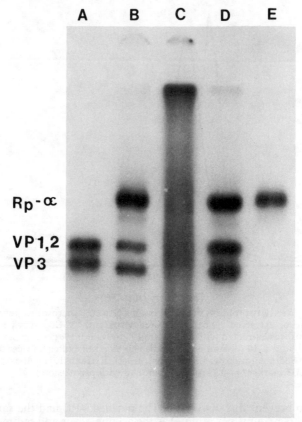

FIG. 2. Autoradiograph of the ^{125}I-labeled CB3 VRC after sodium dodecyl sulfate-polyacrylamide gel electrophoresis. Lane A, 7,500 cpm of ^{125}I-CB3; B, 14,000 cpm of ^{125}I-VRC; C, 40,000 cpm of ^{125}I-labeled control sample; D, 13,000 cpm of ^{125}I-VRC, immunoprecipiated with anti-CB3 serum; and E, 4,500 cpm of ^{125}I-Rp-a isolated from the sucrose gradient in Fig. 3A. On these gels the CB3 VP1 and VP2 comigrate, and VP4 is not seen as a result of poor labeling by ^{125}I. (From reference 10 with permission.)

Rp-a was tested for ability to bind specifically to group B coxsackieviruses. Rp-a (labeled with ^{125}I) was eluted from the VRC (by heating for 2 h at 45°C) and incubated with fresh (unlabeled) viruses for 24 to 48 h at 4°C. These preparations were sedimented on a sucrose gradient, and the 150S region of each gradient was examined for ^{125}I label. The results shown in Fig. 3 demonstrate that Rp-a bound to both CB3 and coxsackievirus B1 (CB1) (Fig. 3A and 3B, respectively), but not to poliovirus T1 (Fig. 3C). The amount of binding to CB3 was greater than to CB1, suggesting that the equilibrium binding constant for the virus and solubilized receptor may differ for these two viruses. When the

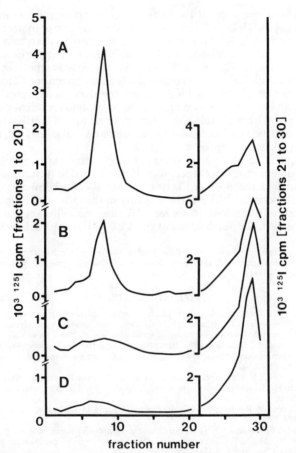

FIG. 3. Radioactive profiles of the fractions of 5 to 30% sucrose gradients centrifuged with samples of ^{125}I-VRC which were heat disrupted and incubated for 48 h at 4°C with 10^9 PFU of fresh, unlabeled virus in phosphate-buffered saline. (A) CB3, (B) CB1, (C) poliovirus, and (D) phosphate-buffered saline as a control. Sedimentation was from right to left. (From reference 10 with permission.)

fresh VRC (CB3) peak was isolated from the sucrose gradient (Fig. 3A), only one iodinated protein, Rp-a, was found by electrophoretic analysis (Fig. 2, lane e). This fresh VRC also was immunoprecipitated by anti-CB3 serum (data not shown).

The cellular receptor site for the group B coxsackieviruses is postulated to be a multicomponent glycoprotein complex with an approximate molecular weight of 275,000 (6). Since Rp-a has a mass of 49,500 daltons, either the receptor site is composed of multiple units of Rp-a or it contains more than one polypeptide chain. We suspect the latter, since a mouse monoclonal antibody prepared against the HeLa cell receptor for CB3 (manuscript in preparation) did not immunoprecipitate Rp-a. Studies are in progress to determine the size of the protein to which the monoclonal antibody binds, but preliminary results suggest that it is larger than Rp-a. We also noted (10) that when the cell extract was treated with sodium dodecyl sulfate the sedimentation coefficient increased from 140S to between 145 and 150S, suggesting that some proteins were lost. The relationship between Rp-a and the receptor protein for adenovirus type 2 is unknown (7, 12). Although these isolated proteins are of similar size (12) and CB3 and adenovirus type 2 compete for receptors on HeLa cells (7), they may be separate proteins in a receptor complex (2).

The technique of affinity purification outlined in this paper may be useful for the identification of other viral receptors. When bound to its natural ligand (the virus), the purified receptor is likely to retain its biochemical activity after purification. The critical factor is the stability of the virus-recpetor bond during extraction and purification. However, if extracted cellular material is not adequately solubilized, purification of the VRC will not be achieved.

This investigation was supported by Public Health Service research grant AI-03771 from the National Institute of Allergy and Infectious Diseases and by institutional Biomedical Research Support grant 2S07RR05413.

LITERATURE CITED

1. **Crowell, R. L.** 1976. Comparative generic characteristics of picornavirus-receptor interactions, p. 179–202. In R. Beers and E. Bassett (ed.), Cell membrane receptors for viruses, antigens, and antibodies, polypeptide hormones, and small molecules. Raven Press, New York.
2. **Crowell, R. L., and B. J. Landau.** 1983. Receptors in the initiation of picornavirus infections, p. 1–42. In H. Fraenkel-Conrat and R. R. Wagner (ed.), Comprehensive virology, vol. 18. Plenum Publishing Corp., New York.
3. **Crowell, R. L., and J. S. Siak.** 1978. Receptor for group B coxsackieviruses: characterization and extraction from HeLa cell membranes, p. 39–53. In M. Polack (ed.), Perspectives in virology. Raven Press, New York.
4. **Hennache, B., and P. Boulanger.** 1977. Biochemical study of KB-cell receptor for adenovirus. Biochem. J. **166**:237–247.
5. **Jacobs, S., and P. Cuatrecasas.** 1981. Affinity chromatography for membrane receptor purification, p. 61–86. In S. Jacobs and P. Cuatrecasas (ed.), Membrane receptors: methods for purification and characterization. Chapman and Hall, New York.
6. **Krah, D. L., and R. L. Crowell.** 1985. Properties of the deoxycholate-solubilized HeLa cell plasma membrane receptor for binding group B coxsackieviruses. J. Virol. **53**:867–870.
7. **Lonberg-Holm, K., R. L. Crowell, and L. Philipson.** 1976. Unrelated animal viruses share receptors. Nature (London) **259**:679–681.
8. **Lonberg-Holm, K., L. B. Gosser, and J. C. Kauer.** 1975. Early alteration of poliovirus in infected cells and its specific inhibition. J. Gen. Virol. **27**:329–342.
9. **Lonberg-Holm, K., and L. Philipson.** 1974. Early interactions between animal viruses and cells. Monogr. Virol. **9**:1–148.
10. **Mapoles, J. E., D. L. Krah, and R. L. Crowell.** 1985. Purification of a HeLa cell receptor protein for the group B coxsackieviruses. J. Virol. **55**:560–566.
11. **Strosberg, A. D.** 1984. Purification of plasma membrane proteins by affinity chromatography, p. 1–13. In J. Venter and L. Harrison (ed.), Receptor purification procedures. Alan R. Liss, Inc., New York.
12. **Svensson, U., R. Persson, and E. Everitt.** 1981. Virus-receptor interaction in the adenovirus system. I. Identification of virion attachment proteins of the HeLa cell plasma membrane. J. Virol. **38**:70–81.

Characterization of the Cellular Receptor Specific for Attachment of Most Human Rhinovirus Serotypes

RICHARD J. COLONNO, JOANNE E. TOMASSINI, PIA L. CALLAHAN, AND WILLIAM J. LONG

Virus and Cell Biology Research, Merck Sharp & Dohme Research Laboratories, West Point, Pennsylvania 19486

Human rhinoviruses (HRVs) are members of the picornavirus family and are the major causative agent of the common cold in humans. Reciprocal competition binding studies have demonstrated that 20 of 24 serotypes of rhinoviruses compete for a single cellular receptor on HeLa cells. Using HeLa cells as an immunogen, we isolated a mouse monoclonal antibody which had the precise specificity predicted by the biological binding study. The receptor antibody is able to protect HeLa cells from infection by 78 of 88 HRV serotypes assayed. Pretreatment of cells with the antibody is also able to protect HeLa cells from infection by coxsackieviruses A11, A18, and A21, while showing no protective effect against a wide range of other viruses. The antibody demonstrates the same tissue tropism as the major group of HRVs, since it binds only to cells of human origin (with one exception, chimpanzee). Solubilization of HeLa cell membranes with detergents and subsequent chromatography on an immune affinity column resulted in the isolation of a 90,000-molecular-weight protein believed to be the major HRV cellular receptor protein.

Picornaviruses attach to animal cells by means of specific receptors on the cell surface (4, 7, 16). Frequently, the possession of appropriate cellular receptors is sufficient to determine the host range for a particular virus (6). Recent reviews (4, 11) of receptors utilized during picornavirus infections indicate that distinct receptor families exist for attachment of specific groups of picornaviruses. Receptor specificity has been further demonstrated by the isolation of monoclonal antibodies which specifically block the attachment of polioviruses (14) and group B coxsackieviruses (2). The human rhinovirus (HRV) family is composed of over 100 antigenically distinct serotypes and offers a unique opportunity to study receptor specificity (12).

HOW MANY HRV RECEPTORS ARE THERE?

This investigation was initiated to establish whether the large group of medically important viruses classified as HRV attach to susceptible cells by a similarly large variety of cellular receptors. Utilizing 10 HRV serotypes, Lonberg-Holm and his colleagues (9, 10) suggested that at least two different

109

HRV receptors exist on HeLa cells. Further evidence for the existence of multiple HRV receptors was the ability of a limited number of HRV serotypes to attach to mouse-derived cells (15). Since the number of HRV serotypes examined in these earlier studies was limited, 24 randomly selected HRV serotypes were used in reciprocal competition binding assays to determine the total number of receptors utilized (1). The results from these assays are summarized in Fig. 1 and clearly define only two receptor groups. The first group contained 20 of the 24 serotypes which competed with one another for a single cellular receptor. A minor group, containing only 4 of the 24 HRV serotypes tested, shared a second and different cellular receptor. Members of this minor receptor group are also capable of binding to mouse L cells while the major group of viruses cannot (1).

Monoclonal antibodies were pursued to further explore the HRV cellular receptors. BALB/c mice were immunized with HeLa cells, and their spleen cells were fused to mouse myeloma cells. The resulting hybridoma cell supernatants were screened in a cell protection assay for their ability to protect cell monolayers from infection by HRVs (3). Of the thousands assayed, only one hybridoma cell culture supernatant was identified which could protect cells from HRV-14 infection. As a primary test of the monoclonal antibody's specificity, HeLa cells were treated with the antibody and challenged with each of the 24 serotypes represented in Fig. 1. The receptor monoclonal antibody appeared to have the precise specificity predicted by the biological study, since only the 20 HRV serotypes belonging to the major receptor group were unable to infect cells in the presence of the antibody (Table 1). The antibody was unable to protect cells from infection by the minor group serotypes, 1A, 2, 44, and 49.

Additional cell protection studies (Table 1) showed that the receptor antibody could not protect cells from infection by other picornaviruses or several other RNA and DNA viruses. Protection was observed in HeLa cells against infection by group A coxsackievirus serotypes 13, 18, and 21. This was not unexpected, since earlier studies (9) showed that coxsackievirus A21 and HRV-14 (major receptor group) competed for the same cellular receptor.

As a final test of the specificity of this receptor antibody, we surveyed 27 cell lines to determine whether the antibody demonstrated the same species tropism as the major group of HRV serotypes. Experiments involved parallel binding studies of radiolabeled HRV-15 and receptor monoclonal antibody. Results showed that the antibody binds only to cell lines to which the virus is capable of binding (3). With the exception of chimpanzee liver cells, specific binding occurred only to cells of human origin. Thus, the cellular receptor required for binding of the antibody to cell surfaces followed the same species tropism as the virus itself.

Convinced that the receptor monoclonal antibody was of the correct specificity, we assayed an additional 64 HRV serotypes against the antibody blockage. Results confirmed our original prediction that almost 90% of all HRV serotypes compete for a single receptor on cells, since the antibody was able to block infection by 78 of the 88 serotypes tested (Table 1). These data clearly confirmed that the ligand structure present on the viral capsid is highly conserved among HRV serotypes and represents the first defined structural link to this group of viruses.

Blocking Serotypes

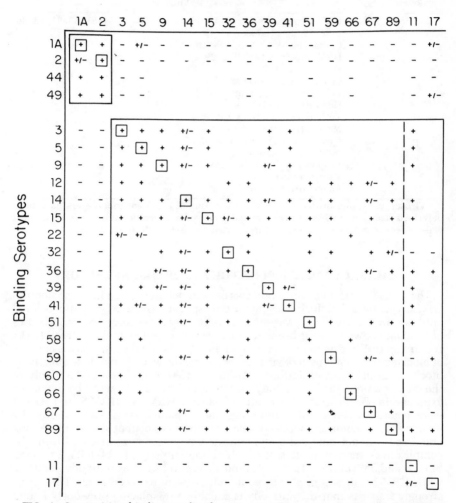

FIG. 1. Competition binding studies between 24 HRV serotypes. Twenty-four HRV serotypes were grown in HeLa R19 cells in the presence or absence of [^{35}S]methionine, the HRVs were purified, and titers were determined as previously described (1). Predetermined amounts of unlabeled virions of the serotypes listed across the top were used to saturate 2×10^5 to 3×10^5 HeLa cells prior to a challenge binding by the ^{35}S-labeled HRV serotypes listed on the left side (1). HRV serotypes which blocked at least 90% as effectively as the homologous virus were scored positive (+), those with which blocking was lower than 20% were scored negative (−), and intermediate percentages were scored +/− to indicate equivocal results. Boxed results are homologous competitions.

TABLE 1. Cell protection assay of receptor monoclonal antibody[a]

Result	Virus
Protection	HRV serotypes 3–28, 32–43, 45, 46, 48, 50–52, 54–61, 63–81, 83–86, 88, 89, and Hanks
	Coxsackieviruses A13, A18, and A21
No protection	HRV serotypes 1A, 1B, 2, 29–31, 44, 47, 49, and 62
	Polioviruses 1, 2, and 3
	Coxsackieviruses B2 and B3
	Echoviruses 1, 3, 6, and 20
	Hepatitis virus type A
	Vesicular stomatitis virus
	Newcastle disease virus
	Parainfluenza viruses 1, 2, and 3
	Influenza A virus
	Respiratory syncytial virus
	Vaccinia virus
	Adenoviruses 1 and 2

[a] Viruses, cells, and procedures used for infection of confluent monolayers have been described elsewhere (3). Assays were scored positive for protection if no cytopathic effect was evident in antibody-treated cultures versus untreated control cultures.

CHARACTERIZATION OF THE MAJOR HRV RECEPTOR

The mouse monoclonal antibody was determined to have an immunoglobulin G1 isotype by gel diffusion using isotype-specific antisera. The monoclonal antibody appears to have a very strong avidity for the receptor protein. Kinetics of binding showed that it binds to the receptor within 1 min, whereas HRVs require up to 1 h to achieve maximal binding (3). While the antibody is capable of blocking the attachment of virus to receptors, the virus is unable to effectively block the attachment of antibody. Although size differences may be involved, the greater affinity of the antibody for the receptor site appears to be the major explanation. To illustrate this point, ^{35}S-labeled HRV-14, HRV-15, and HRV-2 were bound to HeLa cell membranes and unbound virus was removed by washing. The membrane-virus complexes were then incubated with increasing amounts of monoclonal antibody, and the amount of virus released from the complex was measured. Results showed that up to 80% of HRV-14 (weak binder) and 35% of HRV-15 (strong binder) were released by antibody addition, while HRV-2 (minor group) showed no release from its receptor (3). This result strongly suggests that the antibody is capable of displacing previously bound virions of the major group of HRVs and will be a valuable tool in future studies on viral entry.

Since viruses have the capacity to mutate quite rapidly during propagation, we attempted to isolate viral mutants capable of overcoming the receptor antibody block. We first examined our viral stocks for the presence of natural variants. Antibody-treated and untreated confluent monolayers of HeLa cells were infected with 10^9 PFU (multiplicity of infection > 1,000) of HRV-2, HRV-14, HRV-15, and HRV-36, and plates were overlaid with medium (with or without antibody) and 0.4% agar to restrict viral spread. After 3 days of incubation at 34°C, plates were stained to visualize the formation of plaques. The plates containing no antibody showed complete destruction of the cell monolayer by each of the four serotypes, as expected with such a massive amount of virus. However, not a single plaque was observed on the antibody-treated plates infected by HRV-14, HRV-15, and HRV-36 (3). The antibody-treated plate that was infected with HRV-2 was completely destroyed and served as a control. This result demonstrates that natural variants capable of bypassing this specific blockage do not preexist in our concentrated virus preparations.

A second experiment was then designed to select for these mutations. HRV-15 (100 to 200 PFU) was adsorbed to HeLa cell monolayers for 0.5, 2, 4, and 6 h to ensure eclipse of the virus particles prior to the addition of antibody in an agar overlay. After a 3-day incubation, no plaques were present in any of the antibody-treated plates regardless of when the antibody was added (3). While no direct mutagenesis has been attempted, these data strongly suggest that the viral ligand for the major HRV cellular receptor resides in a location that is not easily changed. In addition, this result clearly confirms that HRV transmission occurs only through receptor utilization.

Isolation and identification of the cellular receptor utilized by the major group of HRVs is a prerequisite to complete understanding of viral entry into cells. Solubilization of cellular receptors can be achieved by several methods, of which the most commonly used is detergent treatment. We have used several ionic and nonionic detergents in attempts to solubilize the HRV major receptor and have concluded that 0.3% sodium deoxycholate is the most efficient. Detergent-solubilized receptor preparations were chromatographed on an Affi-gel column coupled to receptor monoclonal antibody (5). After extensive washing, the bound material was eluted from the column with diethylamide and analyzed by polyacrylamide gel electrophoresis. Results (Fig. 2) showed a predominant protein band migrating with an apparent molecular weight of 90,000 (J. E. Tomassini and R. J. Colonno, submitted for publication). No equivalent protein of 90,000 molecular weight was found when Affi-gel columns without antibody were used or when receptors were solubilized from mouse L cells or other cell lines not containing the major HRV receptor. Further analysis of this candidate receptor protein is in progress.

The data presented here illustrate quite conclusively that the vast number (89%) of HRV serotypes utilize a single cellular receptor protein for attachment to susceptible cells. These data were obtained by studying competition for receptors among 24 randomly selected HRV serotypes and by isolation of a mouse monoclonal antibody capable of specifically blocking attachment of this major HRV group. Competition binding studies among the 10 HRV serotypes, belonging to the minor group, have demonstrated that these serotypes compete with one another for a second receptor that is clearly different from the major HRV receptor (3).

Studies with the receptor monoclonal antibody have repeatedly demonstrated the inability of the major group of HRVs to compensate for the receptor block. These results strongly suggest that the viral ligand resides at a location involved in the structural integrity of the capsid. Mutations in this region

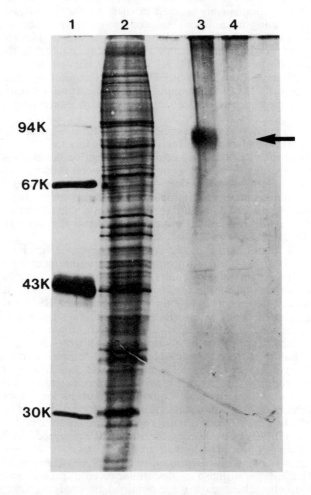

FIG. 2. Affinity chromatography of major HRV cellular receptor proteins. HeLa R19 cell membranes were solubilized with 0.3% sodium deoxycholate, and cellular debris was removed by high-speed centrifugation (Tomassini and Colonno, submitted). Solubilized supernatant was chromatographed on an Affi-gel column or an Affi-gel column containing linked receptor monoclonal antibody (5). The columns were washed extensively in the presence of 0.3% sodium deoxycholate. Bound material was eluted with 0.2% sodium dodecyl sulfate and analyzed on a 10% polyacrylamide-sodium dodecyl sulfate gel (8). The gel was fixed and proteins were visualized by silver staining (13). Lane 1, Protein markers; lane 2, high-speed supernatant; lane 3, Affi-gel with antibody; lane 4, Affi-gel without antibody. Arrow indicates position of receptor protein.

probably result in an instability of the capsid structure and are, therefore, lethal mutations.

While we have isolated a cellular protein involved in HRV attachment, we know little about its normal function or the number of components required to make up an active HRV receptor site. The viral attachment site does not appear to be an essential structure on the HeLa cell surface, since prolonged treatment with the receptor monoclonal antibody results in no inhibition of cell growth or DNA, RNA, and protein synthesis. It is hoped that further characterization and cloning of the receptor protein will provide a better understanding of this important cellular receptor.

LITERATURE CITED

1. **Abraham, G., and R. J. Colonno.** 1984. Many rhinovirus serotypes share the same cellular receptor. J. Virol. **51:**340–345.
2. **Campbell, B. A., and C. E. Cords.** 1983. Monoclonal antibodies that inhibit attachment of group B coxsackieviruses. J. Virol. **48:**561–564.
3. **Colonno, R. J., P. L. Callahan, and W. L. Long.** 1986. Isolation of a monoclonal antibody that blocks attachment of the major group of human rhinoviruses. J. Virol. **57:**7–12.
4. **Crowell, R. L., and B. J. Landau.** 1983. Receptors in the initiation of picornavirus infections, p. 1-42. *In* H. Fraenkel-Conrat and R. R. Wagner (ed.), Comprehensive virology, vol. 18. Plenum Publishing Corp., New York.
5. **DeWitt, D. L., and W. L. Smith.** 1983. Purification of prostacyclin synthase from bovine aorta by immunoaffinity chromatography. J. Biol. Chem. **258:**3285–3293.
6. **Holland, J. J., and B. H. Hoyer.** 1962. Early stages of enterovirus infection. Cold Spring Harbor Symp. Quant. Biol. **27:**101–112.
7. **Krah, D. L., and R. L. Crowell.** 1985. Properties of the deoxycholate-solubilized HeLa cell plasma membrane receptor for binding group B coxsackieviruses. J. Virol. **53:**867–870.
8. **Laemmli, U. K.** 1970. Cleavage of structural proteins during the assembly of the head of bacteriophage T4. Nature (London) **227:**680–685.
9. **Lonberg-Holm, K., R. L. Crowell, and L. Philipson.** 1976. Unrelated animal viruses share receptors. Nature (London) **259:**679–681.
10. **Lonberg-Holm, K., and B. D. Korant.** 1972. Early interaction of rhinoviruses with host cells. J. Virol. **9:**29–40.
11. **Lonberg-Holm, K., and L. Philipson (ed.).** 1981. Receptors and recognition, series B, vol. 8, Virus receptors, part 2. Chapman and Hall, Ltd., London.
12. **Melnick, J. L.** 1980. Taxonomy of viruses. Prog. Med. Virol. **26:**214–232.
13. **Merril, C. R., R. C. Switzer, and M. L. VanKeuren.** 1979. Trace polypeptides in cellular extracts and human body fluids detected by two-dimensional electrophoresis and a highly sensitive silver stain. Proc. Natl. Acad. Sci. USA **76:**4335–4339.
14. **Minor, P. D., P. A. Pipkin, D. Hockley, G. C. Schild, and J. W. Almond.** 1984. Monoclonal antibodies which block cellular receptors of poliovirus. Virus Res. **1:**203–212.
15. **Yin, F. H., and N. B. Lomax.** 1983. Host range mutants of human rhinoviruses in which nonstructural proteins are altered. J. Virol. **48:**410–418.
16. **Zajac, I., and R. L. Crowell.** 1965. Location and regeneration of enterovirus receptors of HeLa cells. J. Bacteriol. **89:**1097–1100.

Encephalomyocarditis Virus Attachment

GRAHAM P. ALLAWAY, INGRID U. PARDOE, AMIR TAVAKKOL,
AND ALFRED T. H. BURNESS

*Faculty of Medicine, Memorial University of Newfoundland, St. John's,
Newfoundland, Canada*

Although the biological significance of the interaction of encephalomyocarditis (EMC) virus and human erythrocytes is unknown, it provides a convenient model for study of the attachment of a virus to receptors. The human erythrocyte receptor is particularly sensitive to neuraminidase, which suggests that it is one or more of the sialoglycoproteins known as glycophorins. Using naturally occurring, specific glycophorin-deficient cells we have obtained evidence that the sole erythrocyte receptor for EMC virus is glycophorin A, which contains 131 amino acids and 16 sialo-oligosaccharide side chains. Comparison with the structure of glycophorin B, which is not a receptor, and protease treatment or chemical modification of normal and deficient cells led to the conclusion that amino acids 35 to 39 (numbered from the N terminus) and the sialo-oligosaccharide linked to threonine at position 37 are particularly important for virus binding to glycophorin A. The polyhydroxy side chain in sialic acid is not required for attachment, but the carboxyl group is. Amino acids 1 to 34 with associated sialo-oligosaccharides are not part of the EMC receptor, nor are gangliosides or other sialoglycoproteins.

Encephalomyocarditis (EMC) virus belongs to the *Cardiovirus* genus of the *Picornaviridae* family (7). In common with other cardioviruses, EMC virus attaches to erythrocytes from different species (1, 18). The biological significance of the interaction between the virus and mammalian erythrocytes, which are incapable of being infected, is unknown, but the intriguing suggestion has been made that such interactions may decrease the susceptibility of the host to infection by aiding clearance of the virus and by presenting the virus more effectively as an antigen to immunocompetent cells (23). Despite our ignorance as to the role of virus-erythrocyte interaction, this interaction provides a convenient model for studying virus attachment and is the subject of this brief review, which comprises three parts. The first part describes studies at the whole-cell level which were aimed at identifying the EMC virus receptor on human erythrocytes. The second part discusses studies at the molecular level that attempted to elucidate the specific region on the receptor involved in virus binding. The third part of the review deals with attachment at the submolecular level. Sialic acid moieties in cellular receptors are required for the attachment of a number of viruses (reviewed in reference 3). The role of sialic acids in the EMC virus-human erythrocyte interaction was examined, and attempts were made to define which groups within the sialic acid residues play a part in the

116

attachment process. In many of the studies reported here the PR-8 strain of influenza virus was used as a control.

IDENTIFICATION OF THE EMC VIRUS RECEPTOR ON HUMAN ERYTHROCYTES

Treatment of erythrocytes with neuraminidase, which releases sialic acid, reduces EMC virus attachment to about 10% of that to untreated cells and completely inhibits hemagglutination (1, 28). This strongly suggests that the EMC virus receptor on erythrocytes contains sialic acid, which is usually found at the free terminus of short, often branched oligosaccharide side chains on glycoconjugates such as glycoproteins and glycolipids (17). Glycoconjugates can be identified by polyacrylamide gel electrophoresis in the presence of sodium dodecyl sulfate followed by periodic acid-Schiff reagent (PAS) staining (12). Examination of human erythrocyte membranes in this way revealed the presence of at least five glycoconjugates, which were originally named PAS 1, 2, 3, and 4 and glycolipid (12). The PAS conjugates are now known to be the dimer (PAS 1) and monomer (PAS 2) forms of glycophorin A (or glycophorin α), the monomer form (PAS 3) of glycophorin B (or glycophorin δ), and a hetero-dimer (PAS 4) comprising one molecule each of glycophorin A and glycophorin B (Fig. 1) (2).

Glycophorins can be extracted from human erythrocyte membranes by a variety of techniques (16). Such preparations inhibit EMC virus attachment and hemagglutination (11). Larger amounts of glycophorins may be examined by sodium dodecyl sulfate-polyacrylamide gel electrophoresis without overloading the gels when isolated rather than complexed in membranes together with an excess of nonglycosylated material. As a result, additional minor components are revealed, including the dimer form of glycophorin B (or glycophorin δ) and monomeric glycophorin C (or glycophorin β) (Fig. 1). Thus, even in the presence of sodium dodecyl sulfate, glycophorins are present as homo- and hetero-dimers, and the purification and biological testing of particular species is no easy task (14).

The problem of aggregation of glycophorins may be overcome by studying these proteins in situ in naturally occurring erythrocyte variants deficient in one or more of the glycophorin species. For instance, En(a−) cells are completely devoid of glycophorin A (Fig. 1), but are otherwise hematologically normal (2, 25). Attachment of ^{3}H-amino acid-labeled EMC virus to such cells was about 8% of that to normal controls, a value similar to that found for attachment to desialylated cells and assumed to be the background value (Table 1). In contrast, ^{32}P-labeled influenza virus attached equally well to En(a−) and normal cells.

The inability of EMC virus to attach to cells which, apart from completely lacking glycophorin A, appear to have the normal complement of glycolipids and other glycophorins (2) suggests that glycophorin A is the sole human erythrocyte receptor for this virus. An alternative explanation is that glycophorins A and B together constitute the EMC virus receptor, and if either is missing, attachment cannot take place. This possibility was tested by using S−s−U+ erythrocytes, which were reported to contain less than 5% of the normal amount of glycophorin B (9). In the cells available to us the amount of

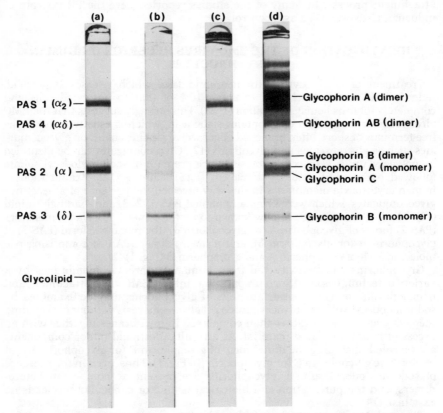

FIG. 1. Sodium dodecyl sulfate-polyacrylamide gel electrophoresis of erythrocytye membranes from (a) normal cells, (b) En(a−) cells, (c) S−s−U+ cells, and (d) a glycophorin preparation. The electrophoresis conditions (19) and staining with PAS reagent (12) were as previously described.

glycophorin B was below the limits of detection under the conditions used (Fig. 1), but nevertheless both ^{3}H-EMC and ^{32}P-influenza virus attached to these cells as efficiently as to normal cells (Table 1). These results demonstrate that glycophorin B is not essential for attachment of EMC or influenza virus and, when considered with the En(a−) cell binding studies above, strongly suggest that only glycophorin A, and none of the minor glycophorins or glycolipids, is involved in EMC virus attachment.

ATTACHMENT SITE FOR EMC VIRUS ON GLYCOPHORIN A

Once the identity of the receptor for EMC virus on human erythrocytes was established, attention could be directed toward elucidating the attachment site

TABLE 1. Attachment of virus to erythrocyte variants and to protease-treated erythrocytes

Cell type	Attachment (%)[a]		Sialic acid on membrane (%)[a]
	EMC virus	Influenza virus	
En(a−)	8	100	34.0
S−s−U+	100	100	ND[b]
Chymotrypsin treated	104	121	50.0
Trypsin treated	49	83	37.7
Ficin treated	2	16	12.4

[a] Normal cells = 100%.
[b] ND, Not determined.

on the receptor. Information can be gleaned from two sources: (i) by comparing the structures of glycophorins A and B, and (ii) by specifically cleaving glycophorin A by use of a variety of proteases and relating loss of various parts of the molecule to changes in biological activity.

Comparison of glycophorin structures. Glycophorin A contains 131 amino acids with 1 N-linked and 15 O-linked sialo-oligosaccharide side chains (26). Approximately 70 of the amino acids in the N-terminal third of the molecule are external to the erythrocyte surface membrane, about 22 lie within the lipid bilayer, and the remaining 39 amino acids are within the cell (Fig. 2). Glycophorin A exhibits polymorphism with serine and glycine in positions 1 and 5 (numbered from the amino terminus) for glycophorin A^M and with leucine and glutamic acid in these positions in glycophorin A^N (15). Glycophorins A^M and A^N have M and N blood group activity, respectively (2).

Glycophorin B is identical to glycophorin A^N in amino acid sequence and oligosaccharide structure for the region of amino acids 1 to 26, except that the complex N-linked oligosaccharide on asparagine 26 is absent (15). Furthermore, amino acids 27 to 35 in glycophorin B are closely homologous to amino acids 56 to 66 in glycophorin A (M or N variety) (8). Thus, glycophorins A and B appear to be nearly identical except that glycophorin B lacks the segment comprising amino acids 27 to 55 with O-linked oligosaccharides at serine 44 and 50 and threonine 37 and 47 (Fig. 3). Since the results with naturally occurring, glycophorin-deficient variants showed that EMC virus attaches to glycophorin A but not to glycophorin B, it is presumed that the sequence of amino acids 27 to 55 with associated oligosaccharides in glycophorin A are responsible for its ability to serve as an EMC virus receptor. The results are also consistent with the possibility that amino acids 1 to 26 with associated oligosaccharides and amino acids 56 to 70 in glycophorin A are not required for EMC virus attachment.

Effect of proteolytic enzymes on EMC virus attachment to erythrocytes. Chymotrypsin is known to cleave glycophorin A in situ, releasing amino acids 1 to 34 and 12 of the 16 associated sialo-oligosaccharide side chains (Fig. 2). Both EMC virus and influenza virus attach to such chymotrypsin-treated cells as efficiently as or more efficiently than to untreated cells (Table 1) (5). This suggests that the amino acids which remain exposed on the cell after

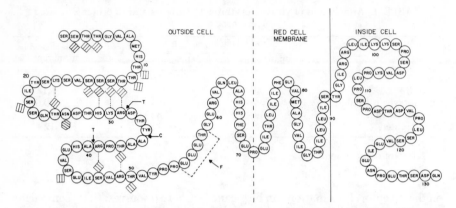

FIG. 2. Amino acid sequence and oligosaccharide attachment sites for glycophorin A^M (modified with permission from reference 22). Symbols: Barred squares, O-linked oligosaccharide; barred hexagon, N-linked oligosaccharide. T, C, and F are trypsin, chymotrypsin, and ficin cleavage sites on intact cells. Dotted lines indicate some possible interactions between positively charged basic amino acids and negatively charged terminal sialic acids on oligosaccharide side chains.

chymotrypsin treatment (alanine 35 to approximately glutamic acid 70) and associated oligosaccharides form part of the attachment site for EMC virus. Although this experiment does not exclude the possibility that amino acids 1 to 34 can themselves serve as a receptor, the above results obtained with En(a−) cells were consistent with no role for amino acids 1 to 26 and their associated sialo-oligosaccharide in the attachment process.

Trypsin cleaves glycophorin A in situ, releasing peptides containing amino acids 1 to 31 and 32 to 39 and associated sialo-oligosaccharide side chains (Fig. 2). Thus, trypsin cleaves glycophorin closer to the membrane than does chymotrypsin, releasing additionally amino acids 35 to 39 with a sialo-oligosaccharide side chain attached to threonine at position 37. Attachment of EMC virus to trypsin-treated cells was about 50% of that to untreated control cells while, in the single measurement made, attachment of ^{32}P-labeled influenza virus to these cells was 83% of that to untreated control cells (Table 1).

The trypsin results suggest that amino acids 35 to 39 and the oligosaccharide side chain linked to threonine 37 are important in EMC virus attachment. They also show that receptor activity is present in the region from amino acid 40 to the erythrocyte surface membrane. The single influenza virus result may be falsely low since trypsin does not cleave glycophorin B (2) which, therefore, should have been available for influenza virus attachment as in the case of En(a−) cells (Table 1).

Ficin cleaves glycophorin closer to the cell surface than does either chymotrypsin or trypsin, but the exact cleavage site has not been defined (Fig. 2 (16). All 16 of the sialo-oligosaccharide side chains on glycophorin A are apparently released since the only PAS-staining material is found in the

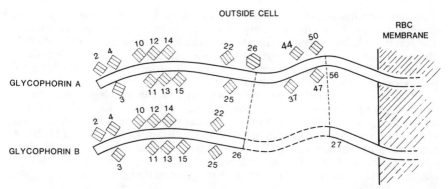

FIG. 3. Diagrammatic representation of glycophorins A and B. Symbols as in Fig. 2. Broken lines indicate the section of glycophorin A which is missing from glycophorin B.

glycolipid region when membranes from ficin-treated cells are examined by sodium dodecyl sulfate-polyacrylamide gel electrophoresis (result not shown). EMC virus attachment to and hemagglutination of ficin-treated cells is destroyed almost completely, whereas influenza virus hemagglutinates such cells, but only about 16% as effectively as untreated cells (Table 1).

ROLE OF SIALIC ACID IN EMC VIRUS ATTACHMENT

The attachment of many unrelated viruses to erythrocytes is prevented if the cells are pretreated with neuraminidase. An obvious explanation of this observation is that all of the viruses concerned attach directly to sialic acid moieties, and when these are removed by neuraminidase, virus attachment is prevented. A proposed alternative hypothesis is that the sialic acid moieties play an indirect role by interacting with basic amino acids to hold the receptor in a particular configuration; in this case, release of sialic acid would destroy the configuration of the receptor and at the same time destroy the structure of each of the different binding sites for the various viruses (3). The hypothesis of an indirect role for sialic acid was similar to that proposed by Lisowska and colleagues (20, 21) to explain why MN blood group activity of glycophorin preparations was destroyed by acetylation, which blocks positively charged lysine residues. We have conducted similar experiments to determine the effect of blocking positively or negatively charged groups in glycophorin on its interaction with virus.

Effect of acetylation of glycophorin on virus attachment. Glycophorin contains six amino groups which can be acetylated under the appropriate conditions (13), one being at the amino terminus and the other five being ε-amino groups in the five lysine residues in the molecule (Fig. 2). Glycophorin preparations in which about five of the six amino groups per molecule had been acetylated were as effective as untreated glycophorin in inhibiting EMC and influenza virus hemagglutination, whereas MN blood group activity was greatly reduced (Table 2). This suggested that neither free ε-amino acids nor the terminal α-amino group was involved in EMC or influenza virus attachment directly or even indirectly, by interacting with sialic acid residues, for instance. The result could have been anticipated for both EMC and influenza viruses since chymotrypsin treatment of erythrocytes releases the only two lysine residues (positions 18 and 30, Fig. 2) exposed on the cell surface and yet attachment of both viruses to these cells is normal or enhanced (Table 1).

Effect of amidation of glycophorin on virus attachment. Negatively charged carboxyl groups, including those in sialic acid residues, can be made neutral by amidation (Fig. 4). When glycophorin preparations were amidated by using glycinamide (6), M blood group activity was substantially depressed (Table 2), confirming an earlier report (10). Similarly, the capacity of amidated glycophorin to inhibit EMC virus hemagglutination was severely reduced while, in contrast, inhibition of influenza virus hemagglutination was unaffected (Table 2).

This result for EMC virus and M blood group activity (but not for influenza virus) is consistent with an indirect role for sialic acid in which amidation blocks its ability to maintain the secondary and perhaps tertiary structure of the receptor. However, the results do not exclude a direct role for sialic acid. For instance, positively charged groups on EMC virus might interact with the negatively charged carboxyl groups in sialic acid, and this might be prevented by amidation. In fact, the results described above, obtained by using erythrocyte variants and studying the effect of proteases on erythrocytes, could be explained in terms of the overall charge on the erythrocyte membrane rather than in terms of the requirement for specific sialic acid or amino acid residues; as the amount of sialic acid on the cell decreases, so does the ability of EMC virus to attach (Table 1).

However, there is evidence to suggest that more than simple negative charges are involved in EMC virus attachment. Previous results showed that controlled neuraminidase treatment of erythrocytes prevents EMC virus attachment or hemagglutination when the amount of sialic acid on the cells falls below about 60% (4). In contrast, chymotrypsin reduces the amount of sialic acid on the cell to below 50% and yet virus attachment is the same on these cells as on untreated cells (Table 1) (4).

If sialic acid residues play an indirect role in the attachment of EMC virus to glycophorin A by interacting with basic amino acids, which amino acids are

TABLE 2. Effect of modifications to glycophorin on hemagglutination (HA) inhibitory properties

Glycophorin modification	HA inhibition (%)[a] against:		
	EMC virus	Influenza virus	Anti-M antibody
Acetylation	100	100	12.8
Amidation	<6	100	24.6
Eight-carbon analog	100	25	50
Seven-carbon analog	100	1.6	25

[a] Untreated glycophorin = 100%.

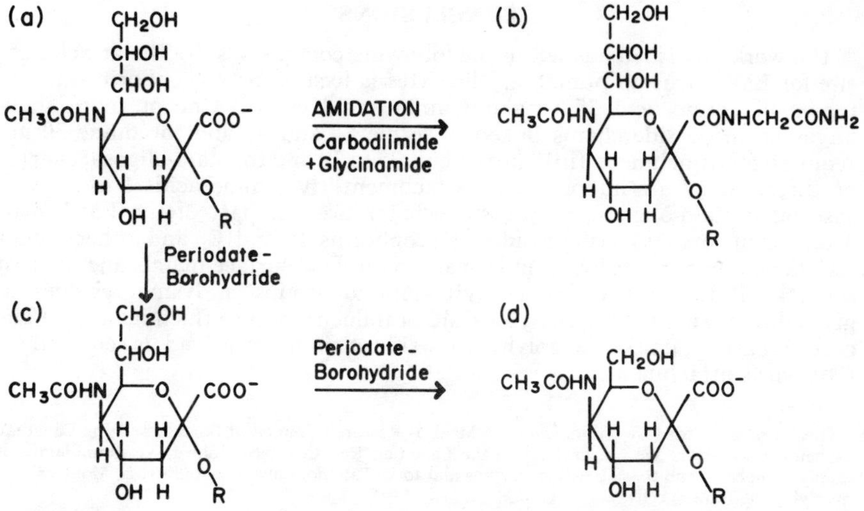

FIG. 4. Structure of N-acetyl neuraminic acid, the form of sialic acid found in human erythrocytes, (a) before and (b) after amidation with glycinamide in the presence of N-ethyl-N'-(3-dimethyl-amino-propyl) carbodiimide hydrochloride (6), or after oxidation with (c) 1 mol, or (d) 2 mol of sodium periodate per mol of sialic acid followed by reduction with sodium borohydride (24). R, Oligosaccharide side chain linking sialic acid to polypeptide.

likely to be involved? The acetylation results above apparently exclude a role for lysine. Arginine residues in positions 39 and 49 and histidine in position 41 (Fig. 2) are alternative candidates. Future experiments planned include treating erythrocytes and glycophorin with butanedione, which blocks positive charges on arginine (29), to determine its effect on virus attachment.

Modification of the polyhydroxy side chain of sialic acid. The sialic acid in human erythrocytes is the N-acetyl neuraminic acid species (27) which contains a polyhydroxy side chain (carbon atoms 7, 8, and 9) (Fig. 4). The side chains of sialic acid residues in glycophorin can be shortened to eight- and seven-carbon analogs by oxidation with molar ratios of periodate to sialic acid of 1 and 2, respectively, followed by borohydride reduction (24).

We confirmed that such modifications of the polyhydroxy side chains reduced both the M blood group activity of glycophorin (10) and its capacity to inhibit influenza virus hemagglutination (24). However, in contrast, the same periodate-oxidized glycophorin preparations inhibited EMC virus hemagglutination as effectively as unoxidized controls (Table 2).

Although the conditions reported by Suttajit and Winzler (24) were used to produce the eight- and seven-carbon analogs, it is still necessary to confirm that the conditions used produced the desired sialic acid derivatives. Tentative conclusions are, however, that while the polyhydroxy side chain in sialic acid is required for influenza virus attachment, it is not necessary for EMC virus attachment.

CONCLUSIONS

The work described has led to the following conclusions. (i) The attachment site for EMC virus on human erythrocytes is located on glycophorin A in the region of amino acid 35 to about amino acid 60. (ii) One or more sialo-oligosaccharide side chains linked to serine 44 and 47 and threonine 50 are required for attachment. (iii) Amino acids 35 to 39 and the sialo-oligosaccharide on threonine 37 are important for attachment. (iv) Amino acids 1 to 34 with associated sialo-oligosaccharide side chains are not part of the EMC virus attachment site. (v) Gangliosides, glycophorins B and C, and other minor sialylated components found in human erythrocyte surface membranes cannot serve as EMC virus receptors. (vi) ε-Amino groups in lysine residues in glycophorin A are not required for EMC or influenza virus attachment. (vii) The carboxyl group, unlike the polyhydroxy side chain in sialic acid, is required for EMC virus attachment.

This work was aided by grants from the Medical Research Council of Canada and the Canadian Diabetes Association. We are grateful to the Canadian Red Cross, St. John's, Newfoundland, for supplying normal and S−s−U+ erythrocytes and to V. Taliano, Canadian Red Cross, Montreal, for supplying the En(a−) cells.

LITERATURE CITED

1. **Angel, M. A., and A. T. H. Burness.** 1977. The attachment of encephalomyocarditis virus to erythrocytes from several animal sources. Virology **82**:428–432.
2. **Anstee, D. J.** 1981. The blood group MNSs-active sialoglycoproteins. Semin. Hematol. **18**:13–31.
3. **Burness, A. T. H.** 1981. Glycophorin and sialylated components as receptors for viruses, p. 63–84. *In* K. Lonberg-Holm and L. Philipson (ed), Virus receptors, part 2. Receptors and recognition. Chapman and Hall, Ltd., London.
4. **Burness, A. T. H., and I. U. Pardoe.** 1981. Effect of enzymes on the attachment of influenza and encephalomyocarditis viruses to erythrocytes. J. Gen. Virol. **55**:275–288.
5. **Burness, A. T. H., and I. U. Pardoe.** 1983. A sialoglycopeptide from human erythrocytes with receptor-like properties for encephalomyocarditis and influenza viruses. J. Gen. Virol. **64**:1137–1148.
6. **Carraway, K. L., and D. E. Koshland.** 1972., Carbodiimide modification of proteins. Methods Enzymol. **25**:616–623.
7. **Cooper, P. D., V. I. Agol, H. L. Bachrach, F. Brown, Y. Ghendon, A. J. Gibbs, J. H. Gillespie, K. Lonberg-Holm, B. Mandel, J. L. Melnick, S. B. Mohanty, R. Povey, R. R. Rueckert, F. L. Schaffer, and D. A. J. Tyrrell.** 1978. Picornaviridae: second report. Intervirology **10**:165–180.
8. **Dahr, W., K. Beyreuther, H. Steinbach, W. Gielen, and J. Krüger.** 1980. Structure of the Ss blood group antigens, II. Hoppe-Seyler's Z. Physiol. Chem. **361**:895–906.
9. **Dahr, W., P. D. Issitt, J. Moulds, and B. G. Pavone.** 1978. Further studies on membrane glycoprotein defects of S-s- and En(a−) erythrocytes. Hoppe-Seyler's Z. Physiol. Chem. **359**:1217–1224.
10. **Ebert, W., J. Metz, and D. Roelcke.** 1972. Modifications of N-acetyl neuraminic acid and their influence on the antigen activity of erythrocyte glycoproteins. Eur. J. Biochem. **27**:470–472.
11. **Enegren, B. J., and A. T. H. Burness.** 1977. Chemical structure of attachment sites for viruses on human erythrocytes. Nature (London) **268**:536–537.
12. **Fairbanks, G., T. L. Steck, and D. F. H. Wallach.** 1971. Electrophoretic analysis of the major polypeptides of the human erythrocyte membrane. Biochemistry **10**:2606–2617.
13. **Fraenkel-Conrat, H.** 1957. Methods for investigating essential groups for enzyme activity. Methods Enzymol. **14**:247–269.
14. **Furthmayr, H.** 1977. Structural analysis of a membrane glycoprotein: glycophorin A. J. Supramol. Struct. **7**:121–134.
15. **Furthmayr, H.** 1978. Structural comparison of glycophorins and immunochemical analysis of genetic variants. Nature (London) **271**:519–524.

16. **Issitt, P. D.** 1981. The MN blood group system. Montgomery Scientific Publications, Cincinnati.
17. **Kornfeld, R., and S. Kornfeld.** 1976. Comparative aspects of glycoprotein structure. Annu. Rev. Biochem. **45:**217–237.
18. **Kunin, C. M.** 1967. Distribution of cell receptors and simple sugar inhibitors during encephalomyocarditis virus-cell union. J. Virol. **1:**274–282.
19. **Laemmli, U. K.** 1970. Cleavage of structural proteins during the assembly of the head of bacteriophage T4. Nature (London) **227:**680–685.
20. **Lisowska, E., and M. Duk.** 1975. Modification of amino groups of human erythrocyte glycoproteins and the new concept on the structural basis of M and N blood group specificity. Eur. J. Biochem. **54:**469–474.
21. **Lisowska, E., and A. Morawiecki.** 1967. The role of free amino groups in the blood group activity of M and N mucoids. Eur. J. Biochem. **3:**237–241.
22. **Marchesi, V. T.** 1979. Functional proteins of the human red blood cell membrane. Semin. Hematol. **16:**3–20.
23. **McClintock, P. C., L. C. Billups, and A. L. Notkins.** 1980. Receptors for encephalomyocarditis virus on murine and human cells. Virology **106:**261–272.
24. **Suttajit, M., and R. J. Winzler.** 1971. Effect of modification of N-acetylneuraminic acid on the binding of glycoproteins to influenza virus and on the susceptibility to cleavage by neuraminidase. J. Biol. Chem. **246:**3398–3404.
25. **Taliano, V., R.-M. Guévin, D. Hébert, G. L. Daniels, P. Tippett, D. J. Anstee, W. J. Mawby, and M. J. A. Tanner.** 1980. The rare phenotype En(a−) in a French-Canadian family. Vox Sang **38:**87–93.
26. **Tomita, M., H. Furthmayr, and V. T. Marchesi.** 1978. Primary structure of human erythrocyte glycophorin A. Isolation and characterization of peptides and complete amino acid sequence. Biochemistry **17:**4756–4770.
27. **Tuppy, H., and A. Gottschalk.** 1972. The structure of sialic acids and their quantitation, p. 403–449. *In* A. Gottschalk (ed.), Glycoproteins. Their composition, structure and function, part A. Elsevier Publishing Co., New York.
28. **Verlinde, J. D., and P. DeBaan.** 1949. Sur l'hemagglutination pas des virus polimoyelitiques murin et la destruction enzymatique de recepteurs de virus poliomyelitique de la cellule receptive. Ann. Inst. Pasteur **77:**632–641.
29. **Yankeelov, J. A.** 1972. Modification of arginine by diketones. Methods Enzymol. **25:**566–579.

Nature of the Interaction between Foot-and-Mouth Disease Virus and Cultured Cells

BARRY BAXT[1] AND DONALD O. MORGAN[2]

Departments of Molecular Biology[1] and Immunology,[2] Plum Island Animal Disease Center, Agricultural Research Service, U.S. Department of Agriculture, Greenport, New York 11944

Studies in our laboratory have concentrated on the nature of both the cellular attachment site and the virion attachment protein for foot-and-mouth disease virus. The virus attaches to cells at a very rapid rate at both 4 and 37°C. The number of cellular attachment sites on cells is in the range of 10^3 to 2.5×10^4 and is dependent on the cell type studied. Saturation binding studies have revealed that six of the seven major serotypes of foot-and-mouth disease virus share at least some common receptor sites which appear to be different from those utilized by poliovirus and encephalomyocarditis virus. Intact functional receptor sites are retained on isolated plasma membranes. Treatment of cells with trypsin abolishes their ability to attach purified virus, and similar treatment of virus decreases its ability to adsorb to cells. Trypsin treatment of virus cleaves the outer virion capsid protein VP_1, suggesting that this protein is the virion attachment protein. Using a series of monoclonal antibodies produced against both intact virus and isolated VP_1, we have identified at least four antigenic areas involved in virus neutralization. One of these areas has been mapped on VP_1 and appears to be responsible for interaction with the cellular attachment site.

Foot-and-mouth disease virus (FMDV), a member of the *Picornaviridae* family, is the cause of foot-and-mouth disease, an economically important, debilitating disease of cloven-footed livestock. The viral capsid consists of 60 copies each of four proteins designated VP_1, VP_2, VP_3, and VP_4. The virus is immunologically complex as evidenced by the fact that there are seven distinct serotypes and greater than 60 subtypes. In addition, the virus undergoes antigenic variation both in nature and in cell culture (44).

Picornaviral infections in cell culture are initiated by the adsorption of virus to specific cell surface receptor sites (CRS) (14, 28), and there is evidence that this interaction is related to viral pathogenicity (15). There have been extensive studies on the early interaction of picornaviruses with cells, yet the membrane macromolecules involved in these interactions have not been identified. Recently, however, a polypeptide of the HeLa cell membrane receptor complex for the group B coxsackieviruses was described (34). A few years ago we began to study these early interactions with FMDV. The purpose of this article is to review results of our studies with intact cells (6, 43), isolated plasma mem-

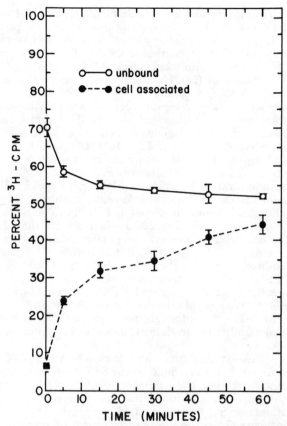

FIG. 1. Adsorption of FMDV to BHK-21 cells. [^{3}H]uridine-labeled FMDV type $A_{12}119$ (10^4 particles per ml) was allowed to adsorb to BHK-21 cells (5×10^7 cells per ml) at 4°C. At various times after addition of virus, the cells were washed with phosphate-buffered saline. Trichloroacetic acid-precipitable counts remaining cell associated and in the washes were determined. (From reference 6. Reprinted with permission.)

branes (7), and the virion attachment protein (8). In addition, we will present some new information on the directions we are taking in this area.

INTERACTION WITH INTACT CELLS

Adsorption. We have investigated the adsorption of six of the seven serotypes of FMDV to both BHK-21 cells and bovine kidney (BK) cells. Our assay system measured the binding of radiolabeled purified virus to cells. A typical binding assay of FMDV type $A_{12}119$ with BHK-21 cells is shown in Fig. 1. Greater than 50% of the total input counts bound rapidly to cells at 4°C within 15 min.

Although the initial binding rate is approximately the same at 4 and 37°C, more virus binds at 4°C (6). The lesser amount of binding at 37°C probably results from the high degree of spontaneous elution from the cell membrane which occurs frequently among picornaviruses (6, 29, 33). The adsorption of FMDV at 4°C is similar to that seen with mengovirus (32), encephalomyocarditis virus (35), and equine rhinovirus (26). However, most human picornaviruses bind very slowly to cells at 4°C (14, 26, 29).

Receptor number. Cellular receptor sites for picornaviruses are finite in number and therefore can be saturated (26, 28, 35). Using a combination of adsorption isotherms and homologous competition binding studies, we determined that there are between 10^4 and 2.5×10^4 CRS per BHK-21 cell for FMDV types $A_{12}119$ and O_{1B} (6). A number within the range of 10^3 to 10^5 CRS per cell is frequently observed for other picornaviruses (28).

Comparisons of saturation by different FMDV serotypes on BK cells have revealed some interesting findings about receptor numbers. Figure 2 shows a saturation binding study comparing types SAT3 and O_{1B} (43). The results show that the binding of type SAT3 decreased when greater than 3×10^3 virus particles per cell were used, whereas it took greater than 6×10^5 virus particles to achieve the same result for type O_{1B}. In addition, 3×10^6 particles of type $A_{12}119$ were required to saturate the CRS on BK cells (43). Thus, there is wide variation of CRS quantity for the different serotypes on BK cells, and there appear to be greater numbers of CRS on BK cells for types O_{1B} and $A_{12}119$ than on BHK-21 cells (6). We have also found that the FMDV SAT serotypes adsorb very poorly to BHK-21 cells, although they replicate efficiently on these cells (43), reflecting an inability to correlate availability of receptors with the ability to replicate.

Recently, we have studied the "ab" antigenic variant of type $A_{12}119$ ($A_{12}119ab$) and found that it is similar to the SAT types in its poor binding to BHK-21 cells and its relatively low numbers of CRS on BK cells (unpublished data). This result offers the possibility that viral antigenic variants might also be receptor variants, as has been postulated previously (15).

Receptor specificity. The saturability of picornavirus receptors has allowed detection of receptors shared by different picornaviruses and led to the concept of receptor "families" (1, 14, 24). We have examined six serotypes of FMDV for receptor specificity and have performed reciprocal competition binding studies between FMDV and two other picornaviruses.

FMDV types $A_{12}119$, O_{1B}, and C_{3Res} appear to utilize some common receptors on BHK-21 cells. However, the lack of total reciprocal binding competition suggested that subsets of receptors may discriminate between these serotypes (6). In BK cells, the SAT serotypes (SAT1, SAT2, and SAT3) utilize the same receptor sites and have some sites in common with types $A_{12}119$ and O_{1B} (43). We have already discussed the differences between the receptor numbers for the SAT types and types $A_{12}119$ and O_{1B} on BK cells. We have also performed competition binding assays between type $A_{12}119$ and encephalomyocarditis virus and poliovirus (43) and have shown that the aphthoviruses utilize a different receptor than these other picornaviruses.

Viral eclipse. After adsorption, the virus undergoes eclipse and uncoating. For enteroviruses and rhinoviruses, uncoating appears to be a multistep process.

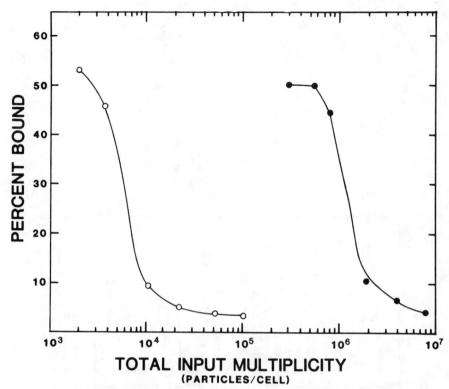

FIG. 2. Saturation binding of types SAT3 and O_{1B}. Increasing input multiplicities of either unlabeled SAT3 (○) or O_{1B}(●) were adsorbed to BK cells for 1.5 h at 4°C, after which [^{3}H]uridine-labeled homologous virus (2×10^3 particles per cell) was added to each tube. Cell-associated trichloroacetic acid-precipitable counts were determined after incubation for 1 h at 4°C. (From reference 43. Reprinted with permission.)

The first step results in an alteration of the viral capsid to form an A particle, which has lost VP_4 and has a reduced sedimentation rate (16, 25–27).

The aphthoviruses appear to undergo eclipse and uncoating as a single step which results in the degradation of 140S virion particles to 12S protein subunits and RNA (Fig. 3) (6, 7, 10). The mechanism of this conversion is not known, but it appears to take place at the plasma membrane (7). Intact FMDV can be broken down to 12S protein subunits by heat, low ionic strength, or acidity. It is possible that the local membrane microenvironment in the vicinity of the receptor might be responsible for this phenomenon, although the action of specific "alteration" proteins has not been ruled out. Recently, it has been reported that lysosomotrophic agents, which raise the pH in endocytic vesicles, inhibit the replication of type O_1 FMDV at an early step in the growth cycle (12).

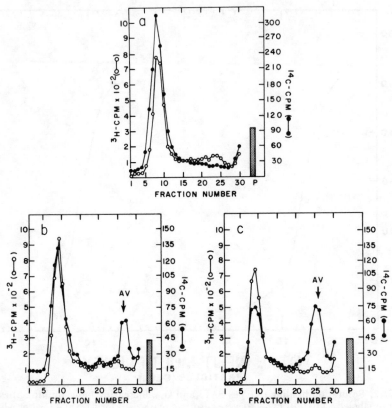

FIG. 3. Analysis of eclipse products. ^{14}C-amino acid-labeled type A_{12}119 (10^3 particles per cell) was adsorbed to BHK-21 cells for 1 h at 4°C. The cells were washed with phosphate-buffered saline, resuspended to 5×10^7 cells per ml, and incubated at 37°C for 0 min (a), 20 min (b), and 50 min (c). Cell homogenates were prepared by Nonidet P-40 treatment and centrifuged on 10 to 50% (wt/vol) sucrose gradients in 20 mM Tris hydrochloride (pH 7.5)–1 M NaCl at 18,000 rpm in an SW41 rotor for 16.5 h at 4°C. The open circles represent a [^{3}H]uridine-labeled marker virus added to each gradient before centrifugation. AV represents the position of acidified [^{35}S]methionine-labeled virus run on a parallel gradient. Sedimentation is from right to left. The pellet fraction, P, is arbitrarily placed to the right of each panel. (From reference 6. Reprinted with permission.)

We have observed similar results for type A_{12} FMDV (B. Baxt, unpublished data), and comparable results have also been obtained for poliovirus (31, 42, 48).

INTERACTION WITH ISOLATED PLASMA MEMBRANES

We have demonstrated that BHK-21 cell plasma membranes contain intact receptors for FMDV type A_{12}119 (7). The kinetics of adsorption to the isolated membranes and intact cells were similar. In both cases binding was inhibited by an excess of unlabeled virus and was greatly reduced at 37°C. If cells were treated with trypsin prior to membrane isolation, the plasma membranes were unable to adsorb virus, suggesting that the virus receptor site is a protein. The isolated membranes were also able to degrade bound virus to 12S protein subunits, as with intact cells.

VIRAL ATTACHMENT PROTEIN

Of the four structural proteins in the FMDV capsid, VP_1 appears to be the outermost protein, as determined by protease sensitivity and iodination of intact virus (7, 39, 46). Trypsin treatment of purified type A FMDV results in the cleavage of VP_1, yielding a 16-kilodalton (kDa) virion-associated fragment (Fig. 4) which spans amino acid residues 1 to 144 (39); a small polypeptide, with a molecular weight less than VP_4, also remains virion associated (unpublished data). Such treatment of type A or O FMDV results in a loss of both infectivity and the ability of the virus to attach to its CRS (4, 7, 11). In addition, purified VP_1 from type A FMDV can elicit neutralizing antibodies and protect animals from virus challenge (2, 3, 22). These data suggest that VP_1 carries the major antigenic determinants of FMDV and is the protein which interacts with CRS on cells. Evidence based on both structural and antigenic studies indicates that VP_1 of other picornaviruses is a major surface protein (19, 23, 30, 45, 47), with the possible exception of coxsackievirus, where VP_2 has been reported to induce neutralizing antibodies (9).

We have examined both neutralization-related and cell attachment epitopes on the surface of type A FMDV with a series of monoclonal antibodies elicited with inactivated intact virus, isolated VP_1, or a 13-kDa CNBr-generated fragment of VP_1 (Table 1) (8, 36). All of the monoclonal antibodies used neutralize viral infectivity; however, studies on the mechanisms of neutralization indicated that these antibodies can be categorized into three groups. The first group, containing antibodies 7SF3, 6HE4, and 2FF11, appeared to cause massive viral aggregation resulting in 50 to 70% inhibition of viral adsorption, with nonspecific residual adsorption (8). A second group, represented by antibody 2PD11, neither aggregated virus nor inhibited binding (Fig. 5c). The third group, represented by antibodies 6HC4 and 6EE2, also did not aggregate virus but inhibited viral adsorption by 80 to 90% (Fig. 5a and b). Epitope mapping of the site on VP_1 which reacts with 6HC4 and 6EE2 placed it at or near the C terminus of the 13-kDa CNBr fragment which ends at amino acid residue 179 (40). Recently, we reported that type A_{12} morphogenic intermediates and acid-derived 12S protein subunits attach specifically to virus receptors (17), suggesting that the cell attachment area on the surface of type A_{12} virus is formed very early in morphogenesis. On the basis of proteolytic enzyme treatment of intact type O_1 virions and VP_1, Strohmaier et al. (46) concluded that the region between residues 138 and 154, which is removed by trypsinization, is necessary for cell binding. Barteling and his co-workers, using a series of trypsin-resistant O_1 mutants (4), suggested that the cell binding site resides in the small fragment of VP_1 which remains virion associated after trypsinization (5). Since at least six of the seven FMDV serotypes share some receptor sites, it is not unreasonable to

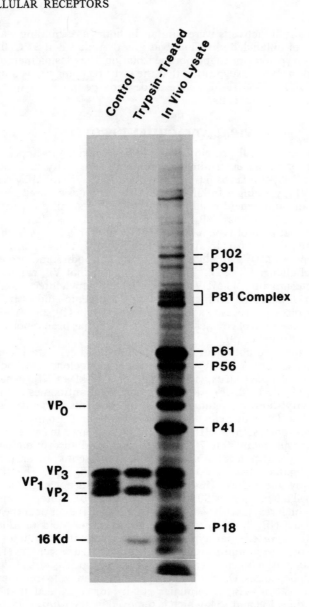

expect that the cell attachment site on the virion would either be conserved in amino acid sequence or have a very similar three-dimensional structure among the serotypes.

FIG. 4. Trypsinization of FMDV. [^{35}S]methionine-labeled type A$_{12}$119ab was treated with purified N-tosyl-L-phenylalanine chloromethyl ketone trypsin (10 U/ml) in TN buffer (10 mM Tris hydrochloride, pH 7.5, 150 mM NaCl) for 15 min at 37°C. The reaction mixture was layered on a 15 to 30% (wt/wt) sucrose gradient in TN buffer and centrifuged at 35,000 rpm for 2.5 h in an SW41 rotor at 4°C. Control virus, incubated without trypsin, was treated similarly. The 140S virus peaks from the gradients were analyzed on a 10% discontinuous sodium dodecyl sulfate-polyacrylamide gel. The in vivo lane is a cell extract of FMDV-infected BK cells labeled with [^{35}S]methionine with the intracellular proteins labeled.

Generation of anti-idiotypic antibodies. Antibodies consist of both variable and constant regions and can themselves act as antigens. Antibodies to the variable region of the immunoglobulin molecules define its idiotype. These anti-idiotypic antibodies may also provide a basis for immune response regulation (20). Recently, anti-idiotypic antibodies have been generated which react with the CRS for reovirus (21, 37, 38) and have been used to isolate a reovirus receptor protein (13).

We have used some of the monoclonal antibodies described in the previous section to inoculate rabbits for producing anti-idiotypic antibodies. The partially purified antibodies reacted only with the monoclonal antibodies which were used to induce them. The only exceptions were the rabbit antisera against monoclonal antibodies 6EE2 and 6HC4. These two monoclonal antibodies have been shown to react with the same or closely situated epitopes (40), and their anti-idiotypic antibodies cross-reacted, indicating that monoclonal antibodies 6HC4 and 6EE2 share an idiotype. These anti-idiotypic antibodies also inhibited the reaction between purified FMDV and the homologous monoclonal antibody, indicating that they reacted at or near the antigen combining site of the inducing monoclonal antibody.

The anti-idiotypic rabbit sera, however, were unable to react specifically with BK cells containing FMDV receptors. In addition, the antisera did not inhibit

TABLE 1. Monoclonal antibodies to type A$_{12}$119ab FMDV^a

Antibody	Antigen^b	Isotype^c	Reactivity^d
2PD11.12.8.1 (2PD11)	Intact virus	IgG2b	V
2FF11.11.4 (2FF11)	Intact virus	IgG3	V, 12S
7SF3.1.H3 (7SF3)	13-kDa	IgG3	V, 12S, 13-kDa, VP$_1$
6HE4.1.1 (6HE4)	VP$_1$	IgG3	V, 12S, 13-kDa, VP$_1$
6FF5	VP$_1$	ND	V, 12S, 13-kDa, VP$_1$
6HC4.1.3 (6HC4)	VP$_1$	IgG2b	V, 12S, 13-kDa, VP$_1$
6EE2.1.2 (6EE2)	VP$_1$	IgG2a	V, 12S, 13-kDa, VP$_1$

^{a} From reference 8.

^{b} Antigens used to elicit the antibodies.

^{c} Determined by agar gel diffusion using commercial isotyping sera (Litton Bionetics). ND, Not determined; Ig, immunoglobulin.

^{d} The ability of the antibody to react with intact virus (V), subunit pentamers of the viral capsid containing VP$_1$, VP$_2$, or VP$_3$ (12S), isolated VP$_1$ (VP$_1$), or the isolated CNBr fragment of VP$_1$ from amino acids 55 to 179 (13-kDa).

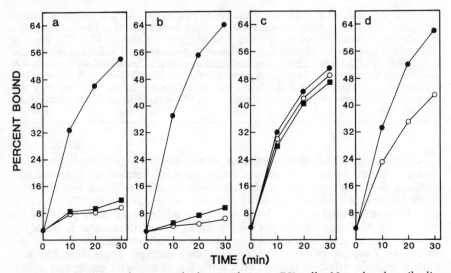

FIG. 5. Adsorption of virus-antibody complexes to BK cells. Monoclonal antibodies were diluted in phosphate-buffered saline with 1% calf serum and mixed with an equal volume of $3H-A_{12}119ab$ diluted 1:135 in the same buffer. The virus-antibody mixture was incubated for 2 h at room temperature. Binding assays were performed on BK cells. (a) Control (●) and 6HC4 at 1:25 (○) and 1:50 (■); (b) control (●) and 6EE2 at 1:25 (○) and 1:50 (■); (c) control (●) and 2PD11 at 1:50 (○) and 1:00 (■); (d) control (●) and 6FF5 at 1:2 (○). (From reference 8.)

the binding of type A_{12} FMDV to BK cells. As a consequence of the very low numbers of CRS on BK cells, our assays may not be sensitive enough to detect antibody binding, or the antibodies may have a low affinity for the receptor sites. Recently, atomic resolution structure analyses for both poliovirus (18) and rhinovirus 14 (41) have revealed the existence of a "canyon" on the icosahedral faces of the virion, suggesting that it is the site for cell receptor binding. Thus, our monoclonal antibodies may not react directly with receptor-specific determinants on the viral surface, making it highly unlikely that the anti-idiotypic antibodies could react with a cell surface receptor.

CONCLUSION

Much progress has been made during the past few years toward understanding the nature of the FMDV-cell interaction; however, the exact nature of the macromolecules in the FMDV receptor site still remains elusive. We have been exploring the viral attachment protein VP_1 and are beginning to define the functional domains of which it is composed. With the application of the techniques of molecular biology and immunology to this problem, the mechanisms surrounding the early events in FMDV-cell interaction should be elucidated within the next few years.

LITERATURE CITED

1. **Abraham, G., and R. J. Colonno.** 1984. Many rhinovirus serotypes share the same cellular receptor. J. Virol. **51:**340–345.
2. **Bachrach, H. L., D. M. Moore, P. D. McKercher, and J. Polatnik.** 1975. Immune and antibody responses to an isolated capsid protein of foot-and-mouth disease virus. J. Immunol. **115:**1636–1641.
3. **Bachrach, H. L., D. O. Morgan, P. D. McKercher, D. M. Moore, and B. H. Robertson.** 1982. Foot-and-mouth disease virus: immunogenicity and structure of fragments derived from capsid protein VP₃ and of virus containing cleaved VP₃. Vet. Microbiol. **7:**85–96.
4. **Barteling, S. J., R. H. Meloen, F. Wagenaar, and A. L. J. Gielkens.** 1979. Isolation and characterization of trypsin-resistant O variants of foot-and-mouth disease virus. J. Gen. Virol. **43:**383–393.
5. **Barteling, S. J., F. Wagenaar, and A. L. J. Gielkens.** 1982. The positively charged structural virus protein (VP₁) of foot-and-mouth disease virus (type O₁) contains a highly basic part which may be involved in early virus-cell interaction. J. Gen. Virol. **62:**357–361.
6. **Baxt, B., and H. L. Bachrach.** 1980. Early interactions of foot-and-mouth disease virus with cultured cells. Virology **104:**42–55.
7. **Baxt, B., and H. L. Bachrach.** 1982. The adsorption and degradation of foot-and-mouth disease virus by isolated BHK-21 cell plasma membranes. Virology **116:**391–405.
8. **Baxt, B., D. O. Morgan, B. H. Robertson, and C. A. Timpone.** 1984. Epitopes on foot-and-mouth disease virus outer capsid protein VP₁ involved in neutralization and cell attachment. J. Virol. **51:**298–305.
9. **Beatrice, S. T., M. G. Katze, B. A. Zajac, and R. L. Crowell.** 1980. Induction of neutralizing antibodies by the coxsackievirus B3 virion polypeptide VP₂. Virology **104:**426–438.
10. **Cavanagh, D., D. J. Rowlands, and F. Brown.** 1978. Early events in the interaction between foot-and-mouth disease virus and primary pig-kidney cells. J. Gen. Virol. **41:**255–264.
11. **Cavanagh, D., D. V. Sanger, D. J. Rowlands, and F. Brown.** 1977. Immunogenic and cell attachment sites of FMDV: further evidence for their location in a single capsid polypeptide. J. Gen. Virol. **35:**149–158.
12. **Carrillo, E. C., C. Giachetti, and R. H. Campos.** 1984. Effect of lysosomotrophic agents on the foot-and-mouth disease virus replication. Virology **135:**542–545.
13. **Co, M. S., G. N. Gaulton, B. N. Fields, and M. I. Greene.** 1985. Isolation and biochemical characterization of the mammalian reovirus type 3 cell-surface receptor. Proc. Natl. Acad. Sci. USA **82:**1494–1498.
14. **Crowell, R. L.** 1976. Comparative generic characteristics of picornavirus-receptor interactions, p. 179–202. *In* R. F. Beers, Jr., and E. G. Basset (ed.), Cell membrane receptors for viruses, antigens and antibodies, polypeptide hormones and small molecules. Raven Press, New York.
15. **Crowell, R. L., B. J. Landau, and J. Siak.** 1981. Picornavirus receptors in pathogenesis, p. 170–180. *In* K. Lonberg-Holm and L. Philipson (ed.), Receptors and recognition. Virus receptors, part 2, Animal viruses, series B, vol. 8. Chapman and Hall, New York.
16. **Crowell, R. L., and L. Philipson.** 1971. Specific alterations of coxsackievirus B3 eluted from HeLa cells. J. Virol. **8:**509–515.
17. **Grubman, M. J., D. O. Morgan, J. Kendall, and B. Baxt.** 1985. Capsid intermediates assembled in a foot-and-mouth disease virus genome RNA-programmed cell free-translation system and in infected cells. J. Virol. **56:**120–126.
18. **Hogle, J. M., M. Chow, and D. J. Filman.** 1985. Three-dimensional structure of poliovirus at 2.9Å resolution. Science **220:**1358–1365.
19. **Hughes, J. V., L. W. Stanton, J. E. Tomassini, W. J. Long, and E. M. Scolnick.** 1984. Neutralizing monoclonal antibodies to hepatitis A virus: partial localization of a neutralizing antigenic site. J. Virol. **52:**465–473.
20. **Jerne, N. K.** 1974. Toward a network theory of the immune system. Ann. Immunol. (Paris) **125C:**373–389.
21. **Kauffman, R. S., J. H. Noseworthy, J. T. Nepom, R. Finberg, B. N. Fields, and M. I. Greene.** 1983. Cell receptors for the mammalian reovirus. II. Monoclonal anti-idiotypic antibody blocks viral binding to cells. J. Immunol. **131:**2539–2541.
22. **Kleid, D. G., D. Yansura, B. Small, D. Dowbenko, D. M. Moore, M. J. Grubman, P. D. McKercher, D. O. Morgan, B. H. Robertson, and H. L. Bachrach.** 1981. Cloned viral protein vaccine for foot-and-mouth disease: responses in cattle and swine. Science **214:**1125–1129.
23. **Lonberg-Holm, K., and B. E. Butterworth.** 1976. Investigation of the structure of polio- and human rhinovirions through the use of selective chemical reactivity. Virology **71:**207–216.

24. **Lonberg-Holm, K., R. L. Crowell, and L. Philipson.** 1976. Unrelated animal viruses share receptors. Nature (London) **259**:679–681.
25. **Lonberg-Holm, K., L. B. Gosser, and J. C. Kauer.** 1975. Early alteration of poliovirus in infected cells and its specific inhibition. J. Gen. Virol. **27**:329–342.
26. **Lonberg-Holm, K., and B. D. Korant.** 1972. Early interaction of rhinovirus with host cells. J. Virol. **9**:29–40.
27. **Lonberg-Holm, K., and L. Philipson.** 1974. Early interaction between animal virus and cells. Monogr. Virol. **9**:1–48.
28. **Lonberg-Holm, K., and L. Philipson.** 1980. Molecular aspects of virus receptors and cell surfaces, p. 789–848. *In* H. M. Blough and J. M. Tiffany (ed.), Cell membranes and viral envelopes, vol. 2. Academic Press, Inc., New York.
29. **Lonberg-Holm, K., and N. M. Whiteley.** 1976. Physical and metabolic requirements for early interaction of poliovirus and human rhinoviruses with HeLa cells. J. Virol. **19**:857–870.
30. **Lund, G. A., B. R. Ziola, A. Salmi, and D. G. Scraba.** 1977. Structure of the mengovirion. V. Distribution of the capsid polypeptides with respect to the surface of the virus particle. Virology **78**:35–44.
31. **Madshus, I. H., S. Olsnes, and K. Sandvig.** 1984. Requirements for entry of poliovirus RNA into cells at low pH. EMBO J. **3**:1945–1950.
32. **Mak, T. W., D. J. O'Callaghan, and J. S. Colter.** 1970. Studies of the early events of the replicative cycle of three variants of Mengo encephalomyelitis virus in mouse fibroblast cells. Virology **42**:1087–1096.
33. **Mandel, B.** 1967. The interaction of neutralized poliovirus with HeLa cells. II. Elution, penetration, uncoating. Virology **31**:248–259.
34. **Mapoles, J. E., D. L. Krah,. and R. L. Crowell.** 1985. Purification of a HeLa cell receptor protein for group B coxsackieviruses. J. Virol. **55**:560–566.
35. **McClintock, P. R., L. C. Billups, and A. L. Notkins.** 1980. Receptors for encephalomyocarditis virus on murine and human cells. Virology **106**:261–272.
36. **Morgan, D. O., B. H. Robertson, D. M. Moore, C. A. Timpone, and P. D. McKercher.** 1983. Aphthoviruses: control of foot-and-mouth disease with genetic engineering vaccines, p. 135–145. *In* E. Kurstak (ed.), Proceedings of the 4th International Conference on Comparative Virology, Banff, Canada, 1982. Marcel Dekker, Inc., New York.
37. **Nepom, J. T., H. L. Weiner, M. A. Dichter, M. Tardieu, D. R. Spriggs, C. F. Gramm, M. L. Powers, B. N. Fields, and M. I. Greene.** 1982. Identification of a hemagglutinin-specific idiotype associated with reovirus recognition shared by lymphoid and neural cells. J. Exp. Med. **155**:155–167.
38. **Noseworthy, J. H., B. N. Fields, M. A. Dichter, C. Sobotka, E. Pizer, L. L. Perry, J. T. Nepom, and M. I. Greene.** 1983. Cell receptors for the mammalian reovirus. I. Syngeneic monoclonal anti-idiotypic antibody identifies a cell surface receptor for reovirus. J. Immunol. **131**:2533–2538.
39. **Robertson, B. H., D. M. Moore, M. J. Grubman, and D. G. Kleid.** 1983. Identification and an exposed region of the immunogenic capsid polypeptide VP₁ on foot-and-mouth disease virus. J. Virol. **46**:311–316.
40. **Robertson, B. H., D. O. Morgan, and D. M. Moore.** 1984. Location of neutralizing epitopes defined by monoclonal antibodies generated against the outer capsid polypeptide, VP₁, of foot-and-mouth disease virus A12. Virus Res. **1**:489–500.
41. **Rossmann, M. G., E. Arnold, J. W. Erickson, E. A. Frankenberger, J. P. Griffith, H. Hecht, J. E. Johnson, G. Kamer, M. Luo, A. G. Mosser, R. R. Rueckert, B. Sherry, and G. Vriend.** 1985. Structure of the human common cold virus and functional relationship to other picornaviruses. Nature (London) **317**:145–153.
42. **Sandvig, K., I. H. Madshus, and S. Olsnes.** 1984. Dimethyl sulfoxide protects cells against polypeptide toxins and poliovirus. Biochem. J. **219**:935–940.
43. **Sekiguchi, K., A. J. Franke, and B. Baxt.** 1982. Competition for cellular receptor sites among selected aphthoviruses. Arch. Virol. **74**:53–64.
44. **Sobrino, F., M. Davila, J. Ortin, and E. Domingo.** 1983. Multiple genetic variants arise in the course of replication of foot-and-mouth disease virus in cell culture. Virology **128**:310–318.
45. **Stanway, G., P. J. Hughes, R. C. Mountford, P. D. Minor, and J. W. Almond.** 1984. The complete nucleotide sequence of a common cold virus: human rhinovirus 14. Nucleic Acids Res. **12**:7859–7875.
46. **Strohmaier, K. R., R. Franze, and K. H. Adam.** 1982. Location and characterization of the antigenic portion of the FMDV immunizing protein. J. Gen. Virol. **59**:295–306.
47. **van der Werf, S., C. Wychowski, P. Bruneau, B. Blondel, R. Crainic, F. Horodniceanu, and M.**

Girard. 1983. Localization of a poliovirus type 1 neutralization epitope in viral capsid polypeptide VP$_1$. Proc. Natl. Acad. Sci. USA **80:**5080–5084.

48. **Zeichhardt, H., K. Wetz, P. Willingmann, and K. O. Habermehl.** 1985. Entry of poliovirus type 1 and mouse Elberfeld (ME) virus into HEp-2 cells: receptor-mediated endocytosis and endosomal or lysosomal uncoating. J. Gen. Virol. **66:**483–492.

Structural Characterization of the Mammalian Reovirus Receptor

MAN SUNG CO,[1,2] GLEN N. GAULTON,[1] JING LIU,[1] BERNARD N. FIELDS,[2] AND MARK I. GREENE[1]

Department of Pathology, Harvard Medical School, Boston, Massachusetts 02115, and Department of Medicine, Tufts University School of Medicine, Boston, Massachusetts 02111,[1] and Department of Microbiology and Molecular Genetics, Harvard Medical School and Shipley Institute of Medicine, and Department of Medicine (Infectious Disease), Brigham and Women's Hospital, Boston, Massachusetts 02115[2]

The mammalian reovirus receptor has been isolated and characterized by using monospecific and monoclonal antireceptor antibodies. The receptor is a protein with a molecular mass of 67 kilodaltons and a pI of 5.8 to 6.0. Neuraminidase treatment reveals a protein of 62 kilodaltons. The reovirus receptor shares many physical features with the mammalian beta-adrenergic receptor, including identical mass and charge. Tryptic peptide patterns are also identical. Monoclonal antireceptor antibodies also inhibit reovirus binding to susceptible tissues in vitro. Furthermore, purified reovirus receptor also binds beta-adrenergic agonists and antagonists.

Cell surface receptors function as specific binding structures for a wide variety of ligands, including antigen, hormones, toxins, and viruses. Viral binding to specific cell surface elements is one of the major determinants of cell and tissue tropism (5, 6). Over the past few years, there have been a number of studies attempting to identify the nature of the cellular membrane proteins that viruses have used as receptors. For example, it has been speculated that Semliki Forest virus binds to histocompatibility antigens in humans and mice (11). Although binding studies have supported this, its significance has been questioned (23) because Semliki Forest virus can bind and subsequently grow in cells devoid of surface histocompatibility antigens. Lactate dehydrogenase virus has been found to interact with mouse Ia antigens, suggesting a role of Ia in targeting virus to a subset of macrophages (13). Rabies virus has been reported to bind in close association with the acetylcholine receptors (18), and Epstein-Barr virus has been shown to recognize the complement receptor type 2 (CR2) of human B lymphocytes (7). More recently, the T-lymphocyte T4 (CD4) antigen has been reported to behave as the receptor for human LAV/HTLV-3/AIDS retrovirus (3, 16).

Work in our laboratory has centered on the utilization of anti-idiotypic (antireceptor) antibodies in the study of the reovirus type 3 cellular receptor. We have used the mammalian reovirus as a model system to determine whether antibodies which recognize and bind reovirus share structural features with nonimmune cell surface reovirus receptors (9). Syngeneic monoclonal and xenogeneic polyclonal anti-idiotypic antibodies were constructed (21, 22) and successfully used to isolate the reovirus receptors on both transformed and

138

normal cells (2). We have further demonstrated that the mammalian reovirus type 3 receptor is similar or identical to the mammalian beta-adrenergic receptor in structure, antigenicity, and ligand binding specificity (2a).

USE OF ANTI-IDIOTYPIC ANTIBODIES AS CELL SURFACE RECEPTOR PROBES

Immunoglobulins can themselves be immunogenic and, when injected into suitable hosts, induce anti-immunoglobulin antibodies which may have novel application. The anti-immunoglobulins of particular interest for study of receptors are those which recognize the ligand binding sites of the inducing antibody molecules. Such anti-immunoglobulins are called anti-idiotypes. Anti-idiotypes have been used to investigate regulatory events which control immune responses in vivo (4, 10). More recently, anti-idiotypic antibodies have been utilized to both identify and isolate a variety of cell surface receptors (8).

The first application of anti-idiotypic antibodies to study surface receptors on cells other than lymphocytes was conducted in 1978 by Sege and Peterson (25). Working directly from assumptions based on Jerne's internal image hypothesis (14), they demonstrated that crude anti-idiotypic antibodies against retinol binding protein-specific immunoglobulins (anti-RBP) bound directly to the RBP binding sites on human prealbumin and on the surface of intestinal epithelial cells (26). This binding was specifically blocked by the addition of anti-RBP antibodies, or by RBP itself.

Sege and Peterson also demonstrated that anti-idiotypic antibodies can mimic the physiological properties of ligand upon binding to cellular receptors (25). They showed that xenogeneic anti-anti-bovine insulin specifically inhibited the binding of ^{125}I-insulin to fat cells. Most significantly, anti-idiotype binding stimulated the uptake of α-amino isobutyric acid in rat cells at levels equivalent to those seen upon insulin treatment. However, the amount of anti-idiotypic antibodies required to induce uptake was 1,000-fold greater (by weight) than that of insulin, which is consistent with the likelihood that the internal image anti-idiotype constitutes only a small fraction of the total polyclonal anti-idiotype repertoire.

Anti-idiotypic antibodies have also been successfully employed in the characterization of the factor H receptor on B lymphocytes (17). The anti-anti-factor H antibodies mimicked factor H by stimulating B cells to release factor I and by potentiating C3b cleavage by factor I in the fluid phase. These anti-idiotypic antibodies were also used in the isolation and characterization of factor H receptor from B cells using immunoprecipitation and sodium dodecyl sulfate-polyacrylamide gel electrophoresis (17). Anti-idiotypes have similarly been used in several studies of the cell surface receptors for adrenergic ligands (12, 24) and formyl peptide chemotaxins (19). In this paper, we report our successful use of anti-idiotypic antibodies in studies of the reovirus type 3 receptor.

ISOLATION AND CHARACTERIZATION OF THE MAMMALIAN REOVIRUS TYPE 3 RECEPTOR USING ANTI-IDIOTYPIC ANTIBODIES

Monoclonal and polyclonal anti-idiotypic antibodies specific for cell surface receptors for mammalian reovirus type 3 have been developed in our laboratories. We started with the reovirus hemagglutinin (HA) which directs tissue binding, tropism, and susceptibility to injury (6). A panel of monoclonal antibodies was prepared in BALB/c mice to the HA protein of reovirus 3 (HA3) (1). One of these monoclonal antibodies (9BG5) was shown to neutralize virus infectivity (1). Rabbit anti-idiotypic antibodies prepared to BALB/c anti-HA3 antibodies were shown to specifically block its ability to bind to purified HA protein (21). More recently, a syngeneic monoclonal anti-idiotype (87.92.6) has been prepared to 9BG5 which displays binding properties identical to those seen with the rabbit anti-idiotypic antibodies (22). The development of the syngeneic monoclonal anti-idiotype allows preparation of large amounts of highly specific antibodies.

We have demonstrated that this monoclonal anti-idiotype and a subset of the polyclonal anti-idiotype antibodies represent an effective "internal image" for the reovirus HA domain that interacts with the cellular receptor for the virus. The evidence supporting this claim is as follows: (i) anti-idiotype binding parallels the cellular tropism of reovirus type 3 (21); (ii) the anti-idiotype and reovirus have similar biological effects in limiting concanavalin A-induced stimulation of murine lymphocytes (20); (iii) the anti-idiotype and reovirus inhibit the host-cell DNA synthesis in a serotype-specific manner (unpublished data); (iv) anti-idiotype specifically inhibits reovirus type 3 binding to target cells (15); (v) anti-idiotype prevents isolated HA protein from binding to monoclonal anti-HA antibodies (21, 22); and (vi) anti-idiotype specifically stimulates both T- and B-cell immunity to reovirus type 3 in vivo (27, 28). On the basis of these observations, we have utilized these anti-idiotypic antibodies as probes to isolate and characterize the cell surface receptors for reovirus type 3.

Our initial experiments indicated that the most effective reovirus receptor purification was achieved by utilizing polyclonal rabbit anti-idiotypic antibodies rather than monoclonal anti-idiotypic antibodies. The R1.1 mouse thymoma cell line, which expresses approximately 70,000 reovirus receptors per cell, has been used as a source of the receptor. R1.1 cells were surface iodinated, and their membranes were solubilized with detergent. The membrane extracts were then incubated with anti-idiotypic antibodies and absorbed with Sepharose-protein A. The proteins bound to protein A were eluted and analyzed by two-dimensional gel electrophoresis under reduced conditions. The major recovered protein had a molecular weight of 67,000 and a heterogeneous pI of 5.8 to 6.0 (Fig. 1A).

Confirmation that the 67,000-dalton structure is the reovirus receptor was achieved by using an immunoblotting technique. R1.1 cell membranes were first purified by differential and equilibrium gradient centrifugation. Membrane extracts were then separated by sodium dodecyl sulfate-polyacrylamide gel electrophoresis under reduced conditions and blotted onto nitrocellulose paper. The reovirus receptor was identified by binding either with ^{125}I-labeled reovirus or with anti-idiotype. Results showed that both radiolabeled virus and anti-idiotype bound to a single band of 67,000 molecular weight (2). Control binding with labeled anti-bovine serum albumin did not detect the band. Thus, recognition of the 67,000-dalton structure is specific for reovirus type 3 particles and anti-idiotypic antibodies. We have also demonstrated directly that the 67,000-dalton receptor on R1.1 is a glycoprotein (2). This was achieved by treatment of immuno-precipitates with neuraminidase, which catalyzes the release of sialic acid residues from glycoproteins, before analysis on sodium dodecyl sulfate-polyacrylamide gels. Neuraminidase treatment generated an additional band of 62,000 molecular weight, presumably a deglycosylated product.

With these anti-idiotypic antibodies we have immunoprecipitated a similar 67,000-dalton structure from a variety of cell lines including rat neuroblastoma, human lymphoma, monkey kidney line, and several other lymphoid lines (2). The presence of a common 67,000-dalton structure on such diverse cell lines and across species strongly suggests that it is associated with normal cellular functions.

Several membrane proteins with molecular weights similar to that of the purified reovirus receptor have been reported. One of these, the mammalian beta-adrenergic receptor, also shows a tissue distribution coincident with that of receptors for virus. To examine the possibility that the mammalian reovirus and beta-adrenergic receptors are related structurally, we obtained a sample of puri-fied calf lung beta-adrenergic receptor and found that the anti-reovirus receptor antibody bound to and precipitated it (2a). Two-dimensional gel analysis of the purified reovirus and beta-adrenergic receptors demonstrated that the two recep-tors are structurally similar, showing indistinguishable molecular weights and isoelectric points (Fig. 1). Trypsin digestion of the two proteins also displayed identical peptide patterns (2a). We have further demonstrated that the purified reovirus receptor binds the beta-antagonists [^{125}I]iodohydroxybenzylpindolol and [^{125}I]iodocyanopindolol in a saturable binding manner, and the binding can be blocked by excess alprenolol and propanolol (unpublished data). All these results suggest that mammalian reovirus may utilize the beta-adrenergic receptor on the host cell for binding and possibly entry. Identification of the cell surface receptor

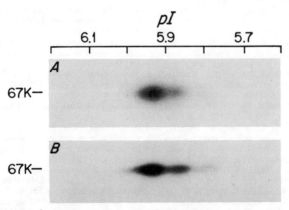

FIG. 1. Vignettes of major spots recovered by two-dimensional gel electrophoresis of immunoprecipitated reovirus type 3 receptor and beta-adrenergic receptor. (A) Reovirus receptor isolated from murine thymoma R1.1 cells. (B) Beta-adrenergic receptor affinity purified from calf lung. Both receptors show a molecular weight of 67,000 and a heterogeneous pI of 5.8 to 6.0. (Reproduced from reference 2a, with permission.)

for the reovirus type 3 is an important step forward in understanding the pathogenesis of reovirus infection.

CONCLUSION

We have summarized the uses of anti-idiotypic antibodies in receptor studies, emphasizing our recent work on the biochemical characterization of the cellular receptors for the mammalian reovirus type 3. Both polyclonal and monoclonal anti-idiotypic antibodies were used as specific probes for the viral receptor. We have isolated the reovirus type 3 receptor and demonstrated its remarkable structural similarities to the beta-adrenergic receptor. These results clearly demonstrate the usefulness of anti-idiotypes in receptor research.

We thank Charles J. Homcy for providing us with the affinity-purified beta-adrenergic receptor from calf lung.

This work was supported by Public Health Service grant PO1-NS16998-04 from the National Institutes of Health.

LITERATURE CITED

1. **Burstin, S. J., D. R. Spriggs, and B. N. Fields.** 1982. Evidence for functional domains on the reovirus type 3 hemagglutinin. Virology **117:**146–155.
2. **Co, M. S., G. N. Gaulton, B. N. Fields, and M. I. Greene.** 1985. Isolation and biochemical characterization of the mammalian reovirus type 3 cell-surface receptor. Proc. Natl. Acad. Sci. USA **82:**1494–1498.
2a. **Co, M. S., G. N. Gaulton, A. Tominaga, C. J. Homcy, B. N. Fields, and M. I. Greene.** 1985. Structural similarities between the mammalian beta-adrenergic and reovirus type 3 receptors. Proc. Natl. Acad. Sci. USA **82:**5315-5318.
3. **Dalgleish, A. G., P. C. Beverley, P. R. Clapham, D. H. Crawford, M. F. Greaves, and R. A. Weiss.** 1984. The CD4 (T4) antigen is an essential component of the receptor for the AIDS retrovirus. Nature (London) **312:**763–767.
4. **Dietz, M. H., M.-S. Sy, B. Benacerraf, A. Nisonoff, M. I. Greene., and R. H. Germain.** 1981. Antigen- and receptor-driven regulatory mechanisms. VII. H-2 restricted anti-idiotypic suppressor factor from efferent suppressor T cells. J. Exp. Med. **153:**450–460.
5. **Dimmock, N. J.** 1982. Initial stages in infection with animal viruses. J. Gen. Virol. **59:**1–22.
6. **Fields, B. N., and M. I. Greene.** 1982. Genetic and molecular mechanisms of viral pathogenesis: implications for prevention and treatment. Nature (London) **300:**19–23.
7. **Fingerorth, J. D., J. J. Weis, J. F. Tedder, J. L. Strominger, P. A. Biro, and P. T. Fearon.** 1984. Epstein-Barr virus receptor of human B lymphocytes is the C3d receptor CR2. Proc. Natl. Acad. Sci. USA **81:**4510–4514.
8. **Gaulton, G. N., M. S. Co, H.-D. Royer, and M. I. Greene.** 1984. Anti-idiotypic antibodies as probes of cell surface receptors. Mol. Cell. Biochem. **65:**5–21.
9. **Greene, M. I., H. L. Weiner, M. Dichter, and B. N. Fields.** 1984. Syngeneic monoclonal anti-idiotypic antibodies identify reovirus type 3 hemagglutinin receptors on immune and neuronal cells, p. 177–187. *In* Monoclonal and anti-idiotypic antibodies: probes for receptor structure and function. Alan R. Liss, Inc., New York.
10. **Hart, D. A., A.-L. Wang, L. L. Pawlak, and A. Nisonoff.** 1972. Suppression of idiotypic specificities in adult mice by administration of anti-idiotypic antibody. J. Exp. Med. **135:**1293–1300.
11. **Helenius, A., B. Morein, E. Fries, K. Simons, P. Robinson, V. Schirrmacher, C. Terhorst, and J. L. Strominger.** 1978. Human (HLA-A and HLA-B) and murine (H2-K and H2-D) histocompatibility antigens are cell surface receptors for Semliki Forest virus. Proc. Natl. Acad. Sci. USA **75:**3846–3850.
12. **Homcy, C. J., S. G. Rockson, and E. Haber.** 1982. An anti-idiotypic antibody that recognized the beta-adrenergic receptor. J. Clin. Invest. **69:**1147–1154.
13. **Inada, T., and C. A. Mims.** 1984. Mouse Ia antigens are receptors for lactate dehydrogenase virus. Nature (London) **309:**59–61.

14. **Jerne, N. K.** 1974. Towards a network theory of the immune system. Ann. Immunol. (Paris) **125c:**373–388.
15. **Kauffman, R. S., J. H. Noseworthy, J. T. Nepom, R. Finberg, B. N. Fields, and M. I. Greene.** 1983. Cell receptors for the mammalian reovirus. II. Monoclonal anti-idiotypic antibody blocks viral binding to cells. J. Immunol. **131:**2539–2541.
16. **Klatzmann, D., E. Champagne, S. Chamaret, J. Gruest, D. Guetard, T. Hercned, J. C. Gluckman, and L. Montagnier.** 1984. T-lymphocyte T4 molecule behaves as the receptor for human retrovirus LAV. Nature (London) **312:**767–768.
17. **Lambris, J. D., and G. D. Ross.** 1982. Characterization of the lymphocyte membrane receptor for factor H (beta₁ H-globulin) with an antibody to anti-factor H idiotype. J. Exp. Med. **155:**1400–1411.
18. **Lentz, T. L., T. G. Burrage, A. L. Smith, J. Crick, and G. H. Tignor.** 1982. Is the acetylcholine receptor a rabies virus receptor? Science **215:**182–184.
19. **Marasco, W. A., and E. L. Becker.** 1982. Anti-idiotype as antibody against the formyl peptide chemotaxis receptor of the neutrophil. J. Immunol. **128:**963–968.
20. **Nepom, J. T., M. Tardieu, R. L. Epstein, J. H. Noseworthy, H. L. Weiner, J. Gentsch, B. N. Fields, and M. I. Greene.** 1982. Virus-binding receptors: similarities to immune receptors as determined by anti-idiotypic antibodies. Surv. Immunol. Res. **1:**255–261.
21. **Nepom, J. T., H. L. Weiner, M. A. Dichter, M. Tardieu, D. R. Spriggs, C. F. Gramm, M. L. Powers, B. N. Fields, and M. I. Greene.** 1982. Identification of a hemagglutinin-specific idiotype associated with reovirus recognition shared by lymphoid and neural cells. J. Exp. Med. **155:**155–167.
22. **Noseworthy, J. H., B. N. Fields, M. A. Dichter, C. Sobotoka, E. Pizer, L. L. Perry, J. T. Nepom, and M. I. Greene.** 1983. Cell receptors for the mammalian reovirus. I. Syngeneic monoclonal anti-idiotypic antibody identifies a cell surface receptor for reovirus. J. Immunol. **131:**2533–2538.
23. **Oldstone, M. B. A., A. Tishon, F. J. Dutko, S. I. T. Kennedy, J. J. Holland, and P. W. Lampert.** 1980. Does the major histocompatibilty complex serve as a specific receptor for Semliki Forest virus? J. Virol. **34:**256–265.
24. **Schreiber, A. B., P. O. Couraud, C. Andre, B. Vray, and D. A. Strosberg.** 1980. Anti-idiotypic antibodies bind to beta-adrenergic receptors and modulate catecholamine-sensitive adenylate cyclase. Proc. Natl. Acad. Sci. USA **77:**7385–7389.
25. **Sege, K., and P. A. Peterson.** 1978. Use of anti-idiotypic antibodies as cell-surface receptor probes. Proc. Natl. Acad. Sci. USA **75:**2443–2447.
26. **Sege, K., and P. A. Peterson.** 1978. Anti-idiotypic antibodies against anti-vitamin A transport protein reacts with prealbumin. Nature (London) **271:**168.
27. **Sharpe, A. H., G. N. Gaulton, H. C. J. Ertl, R. W. Finberg, K. K. McDade, B. N. Fields. and M. I. Greene.** 1985. Cell receptors for the mammalian reovirus. IV. Reovirus-specific cytolytic T cell lines that have idiotypic receptors recognize anti-idiotype B cell hybridomas. J. Immunol. **134:**2702–2706.
28. **Sharpe, A. H., G. N. Gaulton, K. K. McDade, B. N. Fields, and M. I. Greene.** 1984. Syngeneic monoclonal antiidiotype can induce cellular immunity to reovirus. J. Exp. Med. **160:**1195–1205.

Biological Implications of Influenza Virus Receptor Specificity

JAMES C. PAULSON, GARY N. ROGERS, JUN-ICHIRO
MURAYAMA, GLORIA SZE, AND ELAINE MARTIN

Department of Biological Chemistry, University of California-Los Angeles School of Medicine, Los Angeles, California 90024

There is increasing evidence that influenza viruses exhibit differences in receptor specificity. For example, human and avian H3 influenza virus isolates differ in their sialyloligosaccharide receptor specificity, binding preferentially to SAα2,6Gal and SAα2,3Gal linkages, respectively. Receptor variants have been isolated from human and avian isolates which exhibit the opposite specificity of the parent virus. In each case the variants differ from their parent by a single amino acid at residue 226 in the receptor binding pocket of the hemagglutinin (G. N. Rogers, J. C. Paulson, R. S. Daniels, J. J. Skehel, I. A. Wilson, and D. C. Wiley, Nature [London] 304:76–78, 1983). This report examines methods used to assess influenza virus binding specificity and discusses the possible biological consequences of such specificity in the interactions of the viruses with their hosts.

Influenza viruses were first observed to exhibit differences in their receptor binding properties soon after the discovery by Hirst in 1941 that these viruses could agglutinate erythrocytes (14). Indeed, viruses differed in their ability to agglutinate the erythrocytes of different species (5) and differed in their sensitivity to various glycoprotein inhibitors of hemagglutination (4). A major advance in the understanding of the nature of the influenza virus receptor was the recognition that receptor-destroying enzymes found associated with the virus and in the culture filtrates of *Vibrio cholerae* were actually sialidases which prevented viral adsorption by cleaving sialic acid from the cell surface (16). Thus, it was shown that sialic acid is an essential receptor determinant of the virus. Yet the basis for differences in receptor binding properties was still unexplained.

It now appears likely that the variation in receptor binding properties of influenza viruses results in large part from their differential interaction with the variety of sialic acid-containing carbohydrate structures which are found on the cell surface and soluble glycoproteins (7, 17, 20, 23) and on glycolipids (1, 27). The major sources of variation in sialyloligosaccharide structure stem from the diversity of oligosaccharide sequences present on glycoprotein and glycolipid carbohydrate groups (2) and also the type of sialic acid attached. Indeed, the sialic acids are actually a family of over 20 different derivatives of the most common sialic acid, *N*-acetylneuraminic acid (24). The purpose of this review is to briefly describe some of the methods used to assess influenza virus receptor specificity and to discuss the possible biological significance of receptor specificity in vivo.

APPROACHES FOR ANALYSIS OF RECEPTOR SPECIFICITY

For influenza virus, an evaluation of receptor specificity is concerned primarily with recognition of a receptor determinant, or oligosaccharide sequence recognized by the viral binding protein, and to a lesser extent with recognition of which molecules carry the receptor determinants. Indeed, it can be assumed that many cell surface glycoproteins and glycolipids will contain similar if not identical sialyloligosaccharide sequences terminating their carbohydrate groups. Although the influenza virus hemagglutinin can be readily purified, it binds weakly to cells and to most purified glycoproteins and glycolipids. For this reason, assays examining influenza virus receptor specificity take advantage of polyvalent interactions of the intact virus with liposomes containing putative receptor molecules (3, 28) or with intact cells which have been sialidase treated to remove sialic acid and then modified to selectively restore receptors of known structure (1, 7, 20, 23, 27).

Studies with liposomes have mainly shown that both glycoproteins and glycolipids containing sialic acid can mediate viral adsorption (24, 28). In principle, more detailed aspects of receptor specificity could also be examined. For example, liposome preparations containing a series of purified gangliosides could be used to evaluate the sialyloligosaccharide sequence required for adsorption, as has been done for Sendai virus, a paramyxovirus (29).

Sialidase-treated cells, which no longer bind virus, can be used much like liposomes with the potential advantage that they present a more physiological substrate for adsorption (1, 20, 27). For example, sequence-specific restoration of influenza virus adsorption to and penetration of sialidase-treated cultured cells after incorporation of various gangliosides was reported by Bergelson et al. (1). Suzuki et al. examined viral adsorption to sialidase-treated chicken erythrocytes after incorporation of ganglioside GM3 (SAα2,3Galβ1,4GlcNAc-ceramide) containing either N-acetylneuraminic acid or N-glycolylneuraminic acid the sialic acid (27). They were able to show that at low pH the H3 virus A/Aichi/2/68 adsorbed to and fused with erythrocytes containing ganglioside GM3 with N-acetylneuraminic acid, but not GM3 with N-glycolylneuraminic acid. Preferential adsorption to N-acetylneuraminic acid over N-glycolylneuraminic acid was suggested to account for the differential agglutination of the erythrocytes of different species containing these two sialic acids.

We have adopted an approach of examining receptor specificity using sialidase-treated erythrocytes enzymatically modified with a mammalian sialyltransferase to contain a sialyloligosaccharide of defined sequence (20). The general reaction for a sialyltransferase is

CMP-sialic acid + HO-acceptor → sialic acid-O-acceptor + CMP

Each enzyme is specific for the donor substrate, CMP-sialic acid, the terminal sequence of the carbohydrate group which serves as an acceptor substrate, and the anomeric linkage of the glycosidic bond formed in the product. To date, four sialyltransferases have been obtained which form the sequences commonly found in glycoproteins (Fig. 1).

Erythroycte preparations containing each sequence can be readily prepared by resialylating sialidase-treated cells with one of the purified mammalian

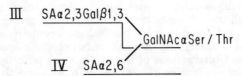

I   SAα2,6Galβ1,4GlcNAcβ1,2Manα1,3

Manβ1,4GlcNAcβ1,4GlcNAcβAsn

II   SAα2,3Galβ1,4(3)GlcNAcβ1,2Manα1,6

III   SAα2,3Galβ1,3

GalNAcαSer / Thr

IV   SAα2,6

FIG. 1. Sialyloligosaccharide sequences of glycoproteins elaborated by purified mammalian sialyltransferases. Structures shown are a typical complex-type oligosaccharide N linked to asparagine and an oligosaccharide O linked to threonine or serine. Underlined sequences labeled I to IV are each produced by a separate sialyltransferase.

sialyltransferases. In this way sialic acid is attached in defined sequence to the carbohydrate groups of endogenous cellular glycoproteins (20). Each erythrocyte preparation can then be tested for its ability to support viral adsorption or hemagglutiation. Comparison of the specificity of over 30 virus strains has revealed that they can differ dramatically in their affinity for binding the sialyloligosaccharide sequences shown in Fig. 1 (13a, 21). Typically, N-acetylneuraminic acid, the most common sialic acid, is incorporated into cells by using CMP-N-acetylneuraminic acid as the donor substrate. However, other sialic acids may also be incorporated by using their corresponding CMP-sialic acid donor substrates (13a). Examples of the differences in receptor specificity detected by hemagglutination of erythrocytes containing the SA2,6Gal(I) or SA2,3Gal(III) linkages with the three most common sialic acids found in mammalian species, N-acetylneuraminic acid, N-glycolylneuraminic acid, or 9-O-acetyl-N-acetylneuraminic acid, are shown for several viruses in Table 1. The biological implications of the differences in receptor specificity detected in this way are described later.

The methods described above for examining receptor specificity of influenza viruses are in most cases also applicable to other viruses which utilize sialyloligosaccharide receptors, such as paramyxoviruses (13, 18, 20), polyoma virus (6), and encephalomyocarditis virus (3). Several other methods of testing virus binding to gangliosides adsorbed to plastic or to thin-layer chromatograms have been used successfully with Sendai virus to show that the sequence SAα2,8SAα2,3Gal- was a high-affinity receptor determinant (12, 15). Such assays may also prove useful for analysis of influenza virus receptor specificity.

CORRELATION OF RECEPTOR SPECIFICITY BASED ON SPECIES OF ORIGIN

One striking result from a screen of influenza viruses was a correlation of receptor specificity with species of origin for isolates with the H3 hemagglutinin

TABLE 1. Hemagglutination (HA) specificities of influenza viruses toward receptor determinants containing N-acetylneuraminic acid (NeuAc), N-glycolylneuraminic acid (NeuGc), and 9-O-acetyl N-acetylneuraminic acid (9-O-Ac-NeuAc)[a]

| Virus | Hemag-glutinin serotype | Viral HA titers of erythrocyte preparations | | | | | |
| | | SAα2,6Gal | | | SAα2,3Gal | | |
		NeuAc	NeuGlc	9-O-Ac-NeuAc	NeuAc	NeuGlc	9-O-Ac-NeuAc
A/RI/5+/57	H2	256	2	2	0	0	0
A/FM/1/47	H1	256	0	0	256	0	0
A/swine/Colorado/1/77	H3	256	256	0	0	0	0
A/duck/Mallard/NY/ 6874/78	H3	256	0	0	512	256	256
A/seal/Massachusetts/ 1/80	H7	0	0	0	256	256	0
A/equine/Miami/1/63	H3	0	0	0	256	0	0

[a] Adapted from Higa et al. (13a).

(21). Human isolates preferentially bound cells containing the SAα2,6Gal linkage but not the SAα2,3Gal linkage, while avian and equine isolates bound cells containing the SAα2,3Gal linkage. Furthermore, human isolates were sensitive to inhibition of hemagglutination or infection by horse serum while avian and equine isolates were relatively insensitive. These findings were of particular interest in view of evidence that the gene coding for the human virus H3 hemagglutinin came from an avian or equine virus (10). Why would receptor specificity differ with species of origin? One attractive possibility is that different host species exert selective pressures resulting in growth to predominance of a receptor-specific virus best suited for growth in that host.

IN VITRO SELECTION OF RECEPTOR-SPECIFIC VARIANTS

For host-mediated selection of receptor-specific influenza viruses, it is envisioned that selection could arise as a consequence of the types of sialyloligosaccharides elaborated on the cell surface or soluble glycoproteins. Given the fact that sialyloligosaccharide structure differs considerably from species to species, such selective pressures would be species specific. Laboratory selection of receptor variants from influenza virus isolates containing the H3 hemagglutinin illustrate two fundamentally different mechanisms.

The first mechanism involves the action of soluble glycoproteins which serve as receptor analogs that bind to the receptor site of the hemagglutinin and block adsorption of the virus to the cell, allowing selection of an inhibitor-insensitive variant. Potent inhibitors of infection in horse serum were first shown to be capable of selecting inhibitor-insensitive variants of human influenza viruses of the H2 serotype (8). Growth of the SAα2,6Gal-specific, inhibitor-sensitive human H3 virus in the presence of horse serum results in selection of variants which exhibit the SAα2,3Gal-specific, inhibitor-insensitive phenotype of the avian and equine isolates (22, 23; see Table 2).

TABLE 2. Summary of receptor specificity and amino acid at residue 226 for influenza viruses with the H3 hemagglutinin[a]

Parent virus	Phenotype	Hemagglutinin specificity		HAI by horse serum	Amino acid at 226
		SAα2,6Gal	SAα2,3Gal		
A/Memphis/102/72					
M1/5	Wild type	64	0	4,096	Leu
M1/HS 10	Wild type	64	0	4,096	Leu
M2	Variant	0	128	128	Gln
M1/HS 8	Variant	0	32	64	Gln
A/Aichi/2/68					
X-31	Wild type	256	0	1,024	Leu
X-31/HS	Variant	0	256	16	Gln
A/duck/Ukraine/1/63					
UK/19	Wild type	0	128	16	Gln
UK/25	Wild type	0	128	32	Gln
UK/43	Variant	128	0	1,024	Leu
UK/49	Variant	128	0	1,024	Leu

[a] Taken in part from Rogers et al. (20a, 22).

The other selection procedure involves selective adsorption of viruses to cell surface receptors. By enzymatic modification of MDCK cells to contain SAα2,3Gal or SAα2,6Gal linkages, it has been possible to show selective propagation of a virus with the complementary specificity from a mixed inoculum containing two viruses of contrasting specificity (S. Carroll and J. Paulson, unpublished data). It has also been possible to select an SAα2,6Gal receptor variant of the SAα2,3Gal-specific avian H3 virus (A/duck/Ukraine/1/63) by repeated rounds of adsorption and elution from SAα2,6Gal modified erythrocytes with intermediate amplification on MDCK cells (20a; see Table 2).

As summarized in Table 2, hemagglutinin sequence analysis of the laboratory-selected variants from three parent viruses has revealed that selection is mediated by a single nucleotide substitution resulting in a shift from leucine to glutamine at amino residue 226, which resides in the receptor binding pocket (20a, 22, 31). Hemagglutinin sequences of field isolates from which the receptor binding properties are known suggest that the same amino acid change is largely responsible for the differences in receptor binding properties of human, avian, and equine isolates containing the H3 hemagglutinin.

HOST-MEDIATED SELECTION OF RECEPTOR VARIANTS

Several examples of the adaptation of influenza viruses to growth in laboratory hosts provide convincing indirect evidence for host-mediated selection of receptor variants of influenza virus. This was first described by Burnet and Bull (5) for changes in the hemagglutination of human, chicken, and guinea pig erythrocytes for human H1 viruses adapted to growth in chicken embryos. More recently, human influenza B viruses adapted to growth in eggs have also been found to exhibit changes in the hemagglutination of the erythrocytes of different species and in their antigenic properties (25). Recombinant viruses constructed by replacing the hemagglutinin of an avian isolate with the H3 hemagglutinin of a human isolate were found to exhibit altered host and tissue tropisms (11, 19). Indeed, the tissue tropism of the recombinants in ducks was correlated with single amino acid substitutions in the hemagglutinin binding site (11).

Recently, we have observed dramatic reversion of the SAα2,6Gal-specific receptor variants of A/duck/Ukraine/1/63 (see Table 2) to the SAα2,3Gal-specific wild-type specificity during a single passage in eggs (20a). This has provided strong evidence for selection mediated at the receptor binding site. However, the mechanism of selection is still not understood. In contrast to the selection exerted on the duck/Ukraine SAα2,6Gal-specific variants, the human SAα2,6Gal-specific viruses grow in eggs with retention of specificity. Despite the similarity in the properties of the SAα2,6Gal-specific viruses stemming from human or avian viruses, further investigation has revealed a number of differences in their receptor binding properties. These include differential hemagglutination of the erythrocytes of different species and human erythrocytes modified to contain the NeuAcα2,6Gal and NeuGcα2,6Gal sequences and differences in their sensitivity to inhibition of plaque formation by horse serum. Further studies suggest that the SAα2,6Gal duck/Ukraine variants grow in chicken embryos, but grow poorly relative to the wild-type virus. Thus, a wild-type revertant rapidly outgrows the variant. The reason for the slow growth of the SAα2,6Gal variants in chicken embryos remains to be established.

Does selection occur in natural hosts? This is the prediction from the correlation of receptor specificity with species of origin for influenza viruses of the H3 serotype (21), but of course this will be difficult to demonstrate directly. In this regard, however, it is of interest that the equine virus A/equine/Miami/1/63 exhibits the inhibitor-insensitive phenotype, and horse serum is a source of a potent inhibitor of infection that results in selection of this phenotype in vitro. Human H3 viruses appear to actively select the SAα2,6Gal-specific, inhibitor-sensitive phenotype. This conclusion is based on the remarkable retention of this receptor specificity in five of the drift strains from 1968 to 1977 and the conservation of leucine at amino acid 226 of the hemagglutinin from 1968 to 1982 (10, 26). It should also be noted that human H2 viruses also appear to have been predominantly the inhibitor-sensitive, SAα2,6Gal-specific phenotype (7–9). This conservation of receptor specificity is in contrast to the spontaneous occurrence of variants of these viruses during passage in laboratory hosts and the relative ease in selecting receptor variants experimentally (23, 30). Ultimately, it will be of interest to understand the degree to which selection of receptor-specific influenza viruses occurs in natural hosts and, where possible, to identify the basis of selection.

LITERATURE CITED

1. **Bergelson, L. D., A. G. Bukrinskaya, N. V. Prokazova, G. I. Shaposhnikova, S. L. Kocharov, V. P. Shevchenko, G. V. Kornilaeva, and E. V. Fomina-Ageeva.** 1982. Role of gangliosides in reception of influenza virus. Eur. J. Biochem. **128:**467–474.
2. **Berger, E. G., E. Buddecke, J. P. Kammerling, A. Kobata, J. C. Paulson, and J. F. G. Vliegenthart.** 1982. Structure biosynthesis and functions of glycoprotein glycans. Experientia **38:**1129–1258.
3. **Burness, A. T. H.** 1981. Glycophorin and sialylated components as receptors for viruses, p. 65–84. *In* K. Lonberg-Holm and L. Philipson (ed.), Virus receptors, part 2, Animal viruses, series B, vol. 8. Chapman & Hall, Ltd., London.

4. **Burnet, F. M.** 1984. Mucins and mucoids in relation to influenza virus action. Inhibition of virus haemagglutination by glandular mucins. J. Exp. Biol. Med. Sci. **26:**371–379.
5. **Burnet, F. M., and D. R. Bull.** 1943. Changes in influenza associated with adaptation to passage in chick embryos. J. Exp. Biol. Med. Sci. **21:**55–69.
6. **Cahan, L. D., R. Singh, and J. C. Paulson.** 1983. Sialyloligosaccharide receptors of binding variants of polyoma virus. Virology **130:**281–289.
7. **Carroll, S. M., H. H. Higa, and J. C. Paulson.** 1981. Different cell surface receptor determinants of antigenically similar influenza virus hemagglutinins. J. Biol. Chem. **256:**8357–8363.
8. **Choppin, P. W., and I. Tamm.** 1960. Studies of two kinds of virus particles which comprise influenza A2 virus strains. Characterization of stable homogeneous substrains in reactions with specific antibody, mucoprotein inhibitors, and erythroyctes. J. Exp. Med. **112:**895–920.
9. **Cohen, A., and G. Belyavin.** 1959. Haemagglutination-inhibition of Asian influenza viruses: a new pattern of response. Virology **7:**59–74.
10. **Fang, R., W. Min Jou, D. Huylebroeck, R. Devos, and W. Fiers.** 1981. Complete structure of A/duck/Ukraine/1/63 influenza hemagglutinin gene: animal virus as progenitor of human H3 Hong Kong 1968 hemagglutinin. Cell **25:**315–323.
11. **Ghendon, Y. Z., A. I. Klimov, and V. P. Ginzburg.** 1984. Studies of a recombinant which inherited the haemagglutinin from the human influenza virus A/Hong Kong/1/68 (H3N2) and other genes from influenza virus A/duck/Ukraine/1/63 (H3N8). J. Gen. Virol. **65:**165–172.
12. **Hansson, G. C., K.-A. Karlsson. G. Larson, N. Stromberg, H. Thurin, C. Orvell, and E. Norrby.** 1984. A novel approach to the study of glycolipid receptors for viruses, binding of Sendai virus to thin-layer chromatograms. FEBS Lett. **170:**15–18.
13. **Haywood, A. M.** 1974. Characteristics of Sendai virus receptors in a model membrane. J.Mol. Biol. **83:**427–436.
13a.**Higa, H. H., G. N. Rogers, and J. C. Paulson.** 1985. Influenza virus hemagglutinins differentiate between receptor determinants bearing N-acetyl-, N-glycollyl-, and N, O-diacetylneuraminic acids. Virology **144:**279-282.
14. **Hirst, G. K.** 1941. Agglutination of red cells by allantoic fluid of chick embryos infected with influenza virus. Science **94:**22–23.
15. **Holmgren, J., L. Svennerholm, H. Elwing, P. Fredman, and O. Strannegard.** 1980. Sendai virus receptor: proposed recognition structure based on binding to plastic-adsorbed gangliosides. Proc. Natl. Acad. Sci. USA **77:**5693–5697.
16. **Klenk, E., H. Faillard, and H. Lempfrid.** 1955. The enzymatic activity of influenza virus. Hoppe-Seyler's Z. Physiol. Chem. **301:**235–246.
17. **Levinson, B., D. Pepper, and G. Belyavin.** 1969. Substituted sialic acid prosthetic groups as determinants of viral hemagglutination. J. Virol. **3:**477–483.
18. **Markwell, M. A. K., L. Svennerholm, and J. C. Paulson.** 1981. Specific gangliosides function as host cell receptors for Sendai virus. Proc. Natl. Acad. Sci. USA **78:**5406–5410.
19. **Naeve, C. W., V. S. Hinshaw, and R. G. Webster.** 1984. Mutations in the hemagglutinin receptor-binding site can change the biological properties of an influenza virus. J. Virol. **51:**567–569.
20. **Paulson, J. C., J. E. Sadler, and R. L. Hill.** 1979. Restoration of specific myxovirus receptors to asialoerythrocytes by incorporation of sialic acid with pure sialyltransferases. J. Biol. Chem. **254:**2120–2124.
20a.**Rogers, G. N., R. S. Daniels, J. J. Skehel, P. C. Wiley, X.-F. Wang, H. H. Higa, and J. C. Paulson.** 1985. Host mediated selection of influenza virus receptor variants: sialic acid-α2-6Gal-specific clones of A/duck/Ukraine/1/63 revert to sialic acid-α2,3Gal-specific wild type in ovo. J. Biol. Chem. **260:**7362-7367.
21. **Rogers, G. N., and J. C. Paulson.** 1983. Receptor determinants of human and animal influenza virus isolates: differences in receptor specificity of the H3 hemagglutinin based on species of origin. Virology **127:**361–373.
22. **Rogers, G. N., J. C. Paulson, R. S. Daniels, J. J. Skehel, I. A. Wilson, and D. C. Wiley.** 1983. Single amino acid substitutions in influenza haemagglutinin change receptor binding specificity. Nature (London) **304:**76–78.
23. **Rogers, G. N., T. J. Pritchett, J. L. Lane, and J. C. Paulson.** 1983. Differential sensitivity of human, avian and equine influenza A viruses to a glycoprotein inhibitor of infection: selection of receptor specific variants. Virology **131:**394–408.
24. **Schauer, R.** 1982. Chemistry, metabolism, and biological functions of sialic acids. Adv. Carbohydr. Chem. Biochem. **40:**131–234.
25. **Schild, G. C., J. S. Oxford, J. C. de Jong, and R. G. Webster.** 1983. Evidence for host-cell selection of influenza virus antigenic variants. Nature (London) **303:**706–709.

26. **Skehel, J. J., R. S. Daniels, A. R. Douglas, and D. C. Wiley.** 1983. Antigenic and amino acids sequence variations in the haemagglutinins of type A influenza viruses recently isolated from human subjects. Bull. WHO **61:**671–676.
27. **Suzuki, Y., M. Matsunaga, and M. Matsumoto.** 1985. N-acetylneuraminyllactosylceramide, $G_{M3-NeuAc}$, a new influenza A virus receptor which mediates the adsorption-fusion process of viral infection: binding specificity of influenza virus A/Aichi/2/68 (H3N2) to membrane-associated G_{M3} with different molecular species of sialic acid. J. Biol. Chem. **260:**1362–1365.
28. **Tiffany, J. M., and H. A. Bough.** 1971. Attachment of myxoviruses to artificial membranes; electron microscopic studies. Virology **44:**18–28.
29. **Umeda, N., S. Nojima, and K. Inoue.** 1984. Activity of human erythroycte gangiosides as a receptor to HVJ. Virology **133:**172–182.
30. **Webster, R. G., W. G. Laver, G. M. Air, and G. C. Schild.** 1982. Molecular mechanisms of variation in influenza viruses. Nature (London) **296:**115–121.
31. **Wilson, I. A., J. J. Skehel, and D. C. Wiley.** 1981. Structure of the haemagglutinin membrane glycoprotein of influenza virus at 3Å resolution. Nature (London) **289:**366–373.

Nature of the Rabies Virus Cellular Receptor

WILLIAM H. WUNNER AND KEVIN J. REAGAN†

The Wistar Institute of Anatomy and Biology, Philadelphia, Pennsylvania 19104

The interaction of rabies virus with putative cellular receptors has been investigated recently in a variety of systems. A specific, saturable receptor for rabies virus was analyzed on cultured cells, and its biochemical nature was investigated by using a soluble octylglucoside extract of cellular membrane which was capable of inhibiting the attachment of radiolabeled rabies virus to BHK-21 cells. The active constituents of the soluble extract were phospholipase and neuraminidase sensitive. Competition studies with other viruses suggest that rabies virus binds to a rhabdovirus-common receptor on cultured cells. Isolated rabies virus glycoprotein has failed to compete in an equivalent manner. The restricted tissue involvement of the highly neurotropic rabies virus in vivo and the location of rabies virus by indirect immunofluorescence at neuromuscular junctions of mouse diaphragm tissue (T. L. Lentz, T. G. Burrage, A. L. Smith, J. Crick, and G. H. Tignor, Science 215:182–184, 1982) has led to the suggestion that the nicotinic acetylcholine receptor complex may serve as a cellular receptor for rabies virus. In testing the generality of this suggestion, we found that the presence of the acetylcholine receptor on the cell surface membrane of various cell lines is not an obligate factor for rabies virus susceptibility of those cells.

The restricted tissue involvement of rabies virus, especially in the early stages of infection and invasion of the central nervous system, is an intriguing aspect of rabies pathogenesis. Electron microscopic observation and immunofluorescence visualization of virus and viral antigen in infected tissue suggest that rabies virus initially replicates only in striated muscle at the site of virus inoculation, if it replicates at all, and does not spread to other tissues of the body before entering localized peripheral nerves (1, 3, 11). There is a strong suggestion that specific elements in nerve endings and synaptic junctions may be involved in the adsorption and uptake of rabies virus. Furthermore, virus spread between neuronal cells in the central nervous system, particularly in spinal and dorsal root ganglia and in the brain, seems to occur by axonal transport along nerve tracts which connect distant nerve centers, by direct cell-to-cell transmission of virus between contiguous cells, and to a lesser extent through intercellular space (7, 11). The last requires that mature virus particles bud from the plasma membrane, and this has been observed in studies of rabies virus-infected mouse brains (4, 6) and brains of two human patients (5). After

†Present address: Glasgow Research Laboratory, E. I. du Pont de Nemours & Co., Inc., Wilmington, DE 19898.

TABLE 1. Effect of polyions on the development of rabies virus antigen or plaques

Polyion added (50 μg/ml)	Immunofluorescence[a]	Plaque assay[b] (PFU/ml)
None	1:30	3.6×10^7
Dextran sulfate	<1:10	ND
DEAE-D	1:150	6.0×10^7

[a] Dilution of strain ERA rabies virus capable of infecting BHK-21 cells in the presence or absence of polyion. The immunofluorescence assay was performed with fluorescein isothiocyanate-labeled antinucleocapsid monoclonal antibody.

[b] Performed with strain CVS rabies virus. ND, Not done.

infecting the central nervous system, rabies virus disseminates from the brain via nerve pathways to other tissues and organs, where it replicates in susceptible nonneural cells. Infection of the salivary glands contributes to the transmission of disease.

Virus receptors on susceptible cells have been considered to be determinants of cellular tropism (2). The importance of cellular receptor sites (CRS) is not well established for rabies virus infection. In this paper we review the role of receptors for rabies virus and discuss future directions of research.

RABIES VIRUS BINDS TO SATURABLE CRS IN VITRO

Whereas rabies virus seems to be a strict neuropathogen in vivo, it infects a wide variety of mammalian and avian cell types in vitro (24). The questions which might be asked, therefore, are (i) is the observed restricted host range of rabies virus in vivo due to an absence of cellular receptors on nonsusceptible cell types, (ii) are receptors expressed only upon tissue cultivation in vitro, and (iii) is the adsorption of rabies virus to cultured cells specific?

Recent studies have demonstrated that rabies virus binds to a specific, saturable site in cells in culture (13, 25), although it has not been proven that virus necessarily enters the cell via these saturable sites. Cultures of mouse neuroblastoma C1300 clone NA (NA) and BHK-21 clone 13 cells have been used to represent cells of neural and nonneural origin. Indeed, the results obtained when NA and BHK-21 cells were used in binding studies were so similar that the distinction of cell origin was unimportant. Initially, it was shown that attachment of fixed rabies virus (ERA and CVS strains) to these cells obeyed the laws of mass action under conditions which were established for optimizing rabies virus binding to cells in cultures (25). The number of virus particles that attached per cell increased linearly as the concentration of virus particles increased. We then compared the kinetics of rabies virus attachment to cells in culture in the presence and absence of polyions, because of an earlier report of the effect of DEAE-dextran (DEAE-D) (8). In accordance with this earlier report, we found that 50 μg of DEAE-D per ml reproducibly enhanced the binding of rabies virus to cells (25), allowed a fivefold increase in viral antigen synthesis as detected by immunofluorescence, and also increased the number and size of plaques (Table 1). In contrast, a polyion of opposite charge, dextran sulfate, at the same concentration enhanced binding of virus to cells (25) but did so

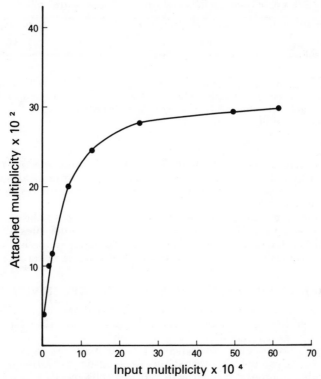

FIG. 1. Saturation of BHK-21 cells with strain ERA rabies virus. A nonsaturable component in binding radiolabeled rabies virus probe to 5×10^6 cells per ml, determined by the binding of probe in the presence of 357 µg of unlabeled strain ERA virus, was subtracted from all values. Taken from Wunner et al. (25).

nonspecifically since viral antigen production was reduced. In the case of DEAE-D, the enhancement of virus attachment was probably due to an effect on the plasma membrane rather than on the virus, since cells pretreated with DEAE-D before virus attachment demonstrated a similar enhanced binding. We did not observe virus aggregation after pretreatment of virus with DEAE-D.

To demonstrate saturability of rabies receptors, cultures of BHK-21 or NA cells were suspended (5×10^6 cells per ml) in a standard assay medium of minimal essential medium buffered to pH 7.4 with 50 mM HEPES (N-2-hydroxyethylpiperazine-N'-2-ethanesulfonic acid), 0.2% bovine serum albumin, and DEAE-D (50 µg/ml), and increasing concentrations of radiolabeled virus were added at 4°C (to prevent penetration) (Fig. 1). The number of virus particles binding to saturable membrane sites was estimated to be from 3×10^3 to 15×10^3 particles per cell (25).

COMPETITION FOR RABIES VIRUS-SPECIFIC CRS IN VITRO

The specificity of the saturable CRS for rabies virus was tested in competition experiments using ^{3}H-labeled ERA virus and excess unlabeled nonpathogenic variant ERA strain virus (RV194-2) or vesicular stomatitis virus. At a concentration of 100μg/ml, these competing viruses caused 70 to 82% inhibition of attachment of radiolabeled virus to NA or BHK-21 cells. The enveloped neurotropic West Nile virus and nonenveloped neurotropic reovirus type 3 did not compete for receptors with labeled rabies virus (25). Purified ^{125}I-labeled rabies virus glycoprotein also failed to compete with unlabeled rabies virus. Perrin and colleagues (12) reported similar results when they attempted to saturate the cellular membrane sites with isolated rabies virus glycoprotein. On the other hand, purified glycoprotein inserted into virosomes prepared from viral lipids effectively competed with rabies virus binding to BHK-21 cells, suggesting that the viral glycoprotein is effective when presented at a suitably high valency (25).

These results indicate that there are specific saturable receptor sites for rabies virus attachment on the surface of cells in culture by the criteria of saturability and competition (22). A third criterion, biologic specificity, requires virus to bind to all cells which may become infected. To date, it has not been possible to find a cell culture that rabies virus does not infect, and there is also often nonspecific rabies virus adherence to the cell surfaces. It may be difficult to demonstrate true "biologic specificity" of receptors for rabies virus.

BIOCHEMICAL NATURE OF THE CRS FOR RABIES VIRUS

Cells were predigested for 30 min at 37°C with trypsin (0.1%), chymotrypsin (0.1%), or neuraminidase (1 U/ml), and in each instance rabies virus attachment was practically unaffected. Superti et al. (19) reported that pretreatment of cells in culture with neuraminidase (from *Clostridium perfringens*) inhibited rabies virus infection. The results with neuraminidase and proteolytic enzymes reported by Superti et al. (19) were compared with the effects of these enzymes on the ability of octyl-β-D-glucopyranoside (OG) extract of cells, prepared as described by Schlegel et al. (16), to compete with host-cell receptors and inhibit rabies virus attachment. The OG preparation was treated with a variety of enzymes or extracted with chloroform-methanol (Table 2). Proteolytic enzymes (trypsin and chymotrypsin at an enzyme-to-substrate ratio of 1:50) did not inhibit the activity of the OG extract, consistent with experiments using intact cells. Unlike the results we obtained with intact cells, however, but in agreement with the results of Superti et al. (19), digestion of the OG extract with neuraminidase totally eliminated its inhibitor activity.

Phospholipase C, and to a lesser extent phospholipases A$_2$ and D, partially reduced the inhibitor activity of the OG extract for rabies virus attachment. The lost activity might be associated with lipid components of the OG extract, and in support of this, the chloroform–methanol-soluble portion of the OG extract was found to contain some attachment inhibitory activity (Table 2). These findings add to a growing body of evidence indicating that lipids participate in the binding of rhabdoviruses to the surface of cell membranes. Schlegel et al.

TABLE 2. Effect of enzymatic pretreatment or lipid extraction on the ability of OG extract to inhibit rabies virus binding to BHK-21 cells[a]

Prepn	Cell-associated cpm[b]	% Inhibition
[35]S-labeled ERA virus	10,194	0
+ OG extract	1,997	80.4
+Chloroform-methanol extract (20 μl) of OG extract	6,161	39.6
+ OG extract, freeze-thawed	1,293	87.3
+ OG extract + neuraminidase	13,811	0
+ OG extract + trypsin	−200[c]	100
+ OG extract + chymotrypsin	840	91.8
+ OG extract + phospholipase A$_2$	3,380	47.2
+ OG extract + phospholipase C	8,502	16.6
+ OG extract + phospholipase D	5,762	43.5

[a] OG extract (99.3 μg of protein) was treated with various enzymes. After incubation at 100°C to inactivate the enzymes, the OG extract was added to 10^6 DEAE-D-pretreated BHK-21 cells per 0.2 ml (final volume) and incubated for 10 min at 4°C before the addition of ca. 5×10^4 cpm of [^{35}S]methionine-labeled strain ERA virus. Cell-associated counts after 2 h at 4°C were determined. Taken from Wunner et al. (25).

[b] The counts per minute binding in the presence of 200 μg of unlabeled ERA virus (5,937 cpm) was considered nonspecifically attached and was subtracted from all values.

[c] Cell-associated counts accounted for less than was considered nonspecifically attached (see footnote b).

(16) showed that the vesicular stomatitis virus binding inhibitor in the OG extract from Vero cells was a phospholipid, possibly phosphatidylserine. Others have implicated lipids and glycolipids in attachment of rabies virus and vesicular stomatitis virus (12, 18, 20). Identification of a specific lipid component(s) and its possible role in the neurotropism of rabies virus in vivo will be an area of investigation in several laboratories, including our own, in the immediate future.

RABIES VIRUS AND THE ACh RECEPTOR

Rabies virus specifically enters neural pathways via motor nerve endings in muscles (23), suggesting the existence of receptors which direct its transport along nerve tracts to the central nervous system. Lentz and co-workers (9) infected isolated murine diaphragms with rabies virus and compared the distribution of inoculum virus with that of high-density clusters of acetylcholine (ACh) receptors at neuromuscular junctions. They also compared the distribution of viral antigen on the surface of cultured chick myotubes with clusters of ACh receptor detected by binding of α-bungarotoxin (αBTX). The morphological appearance of viral antigen patches and ACh receptor clusters at neuromuscular junctions or on cell surfaces showed that the virus and ACh receptors were located close to each other, if not coincident. Furthermore, since these authors were able to inhibit rabies virus infection in the cultured chick myotubes by treatment with αBTX or D-tubocurarine, they suggested that the nicotinic ACh receptor may be the cellular receptor unit for rabies virus.

TABLE 3. Presence or absence of ACh receptor complex and susceptibility to rabies virus[a]

Cell line[b]	Fusion	[125I]αBTX specifically bound[c] (cpm per culture)
BHK-21	−	0
CER	−	120
NA	−	200
SAA	−	46
L8Cl3U	−	0
L8, 24 h	−	0
L8, 216 h	+	29,040

[a] Taken from Reagan and Wunner (14). At 24 h postinfection with 2×10^6 PFU of rabies virus strain CVS per ml, 100% of cells showed immunofluorescent antigen both in the absence and the presence of 5×10^{-4} M D-tubocurarine.

[b] Monolayers (78% confluent) were analyzed in duplicate; 216-h L8 cultures were analyzed when extensive myotube formation was evident.

[c] Amount of [125I]αBTX (450 Ci/mmol) bound in excess of the αBTX bound in the presence of 10^{-4} M D-tubocurarine. This concentration achieved maximum inhibition of 10^{-8} M [125I]αBTX binding to cells with high-density ACh receptors.

We have asked whether only those cells which contain ACh receptor sites can be infected by rabies virus and whether the virion attachment protein of rabies virus recognizes a component which binds ACh or its probes. Our approach was to examine Yaffe's line of rat skeletal muscle cells (L8), which differentiate with increasing cell density from unicellular myoblasts lacking ACh receptor to multinucleated ACh receptor-bearing myotubes (21, 26). Also, a clonal derivative of L8 cells, designated L8Cl3U, which cannot differentiate to myotubes and thus does not develop ACh receptors, was used as a negative control (17). These cells were compared with BHK-21 cells, mouse NA cells, and Singh's *Aedes albopictus* (SAA) clone C6/36 cells for the presence of ACh receptors and for ability to develop rabies virus-specific antigen (14). ACh receptors were probed by the specific binding of [125I]αBTX (21). Table 3 summarizes the results and indicates that cells which acquired ACh receptor (myotubes; 216 h in culture) were no more infectable with rabies virus than undifferentiated L8 cells (24 h in culture) or the L8Cl3U cells. Similarly, although no significant level of ACh receptor was detected on BHK-21, NA, or SAA cells, each was susceptible to infection, although the infectious virus yield from SAA cells was extremely low (unpublished data). To determine whether infection of cells could be inhibited by an antagonist of the ACh receptor, cells were treated with up to 5×10^{-4} M D-tubocurarine before and during incubation with virus. In no case was the percentage of infected cells reduced (15). This point was further emphasized in experiments in which [3H]leucine-labeled rabies virus attachment to BHK-21 and NA cells was tested directly in the presence of αBTX (10^{-8} M to 10^{-5} M). Inhibition was negligible (0 to 10%) compared with 81% inhibition by unlabeled whole virus (15). Moreover, virus infection of cells under low input multiplicities was not affected by the neurotoxins (14). Rabies virus also bound as efficiently to undifferentiating L8 and nondifferentiating L8Cl3U cells as to

BHK-21 cells, as measured by saturable binding to cell surface receptors. The similarity of rabies virus attachment to the L8 and L8Cl3U cells and also to the susceptible BHK-21 cells suggested that functional "receptor" entities existed on these cells in the absence of the ACh receptor (14).

CONCLUSIONS

The dependence of rabies virus on a receptor for its entry into susceptible cells or uptake by nerve endings has been investigated in different experimental systems, all of which have produced interesting but varied results. In cell culture, it was found that rabies virus binds to a specific, saturable receptor on cells and that this receptor has many of the chemical characteristics which have been described for a cellular receptor that specifically binds vesicular stomatitis virus. Saturable binding is, however, one of two independent modes of attachment in virus-cell interactions, and the possibility has not been ruled out that virus bound to nonsaturable sites might become internalized and productively uncoated. Pursuing the suggestion that the ACh receptor may bind rabies virus in vivo (9, 10), we found that rabies virus binding and replication in cultured cells were not inhibited by competitors of ACh which bind to the nicotinic ACh receptor. Therefore, it is possible that rabies virus does not depend on the ACh receptor complex for infection of susceptible cells in vitro and that more than one type of cellular receptor may exist for rabies virus.

This research was supported by Public Health Service grant AI-18562 from the National Institutes of Health and by the Neurovirology/Neuroimmunology Training Program (NS-07180) funded by the National Institutes of Health.

LITERATURE CITED

1. **Charlton, K. M., and G. A. Casey.** 1979. Experimental rabies in skunks. Immunofluorescence light and electron microscopic studies. Lab. Invest. **41:**36–44.
2. **Crowell, R. L., and B. J. Landau.** 1979. Receptors as determinants of cellular tropism in picornavirus infections, p. 1–33. *In* A. G. Bearn and P. W. Choppin (ed.), Receptors and human diseases. Joseph Macey Foundation, New York.
3. **Fekadu, M., and J. H. Shaddock.** 1984. Peripheral distribution of virus in dogs inoculated with two strains of rabies virus. Am. J. Vet. Res. **45:**724–729.
4. **Iwasaki, Y., and H. F. Clark.** 1975. Cell to cell transmission of virus in the central nervous system. II. Experimental rabies in mouse. Lab. Invest. **33:**391–399.
5. **Iwasaki, Y., D. Liu, T. Yamamoto, and H. Konno.** 1985. On the replication and spread of rabies virus in the human central nervous system. J. Neuropathol. Exp. Neurol. **44:**185–195.
6. **Iwasaki, Y., S. Ohtani, and H. F. Clark.** 1975. Maturation of rabies virus by budding from neuronal cell membrane in suckling mouse brain. J. Virol. **15:**1020–1023.
7. **Johnson, R. T., and C. A. Mims.** 1968. Pathogenesis of viral infections of the nervous system. N. Engl. J. Med. **278:**84–92.
8. **Kaplan, M. M., T. J. Wiktor, R. F. Maes, J. B. Campbell, and H. Koprowski.** 1967. Effect of polyions on the infectivity of rabies virus in tissue culture: construction of a single-step growth curve. J. Virol. **1:**145–151.
9. **Lentz, T. L., T. G. Burrage, A. L. Smith, J. Crick, and G. H. Tignor.** 1982. Is the acetylcholine receptor a rabies virus receptor? Science **215:**182–184.
10. **Lentz, T. L., T. G. Burrage, A. L. Smith, and G. H. Tignor.** 1983. The acetylcholine receptor as a cellular receptor for rabies virus. Yale J. Biol. Med. **56:**315–322.
11. **Murphy, F. A.** 1977. Rabies pathogenesis. Brief review. Arch. Virol. **54:**279–297.
12. **Perrin, P., D. Portnoi, and P. Sureau.** 1982. Etude de l'adsorption et de la penetration du virus rabique: interactions avec les cellules BHK21 et des membranes artificielles. Ann. Virol. (Paris) **133E:**403–422.

13. **Reagan, K. J., and W. H. Wunner.** 1984. Early interactions of rabies virus with cell surface receptors, p. 387–392. *In* D. H. L. Bishop and R. L. Compans (ed.), Nonsegmented negative strand viruses. Academic Press, Inc., San Diego.
14. **Reagan, K. J., and W. H. Wunner.** 1985. Rabies virus interaction with various cell lines is independent of the acetylcholine receptor. Arch. Virol. **84:**277–282.
15. **Reagan, K. J., and W. H. Wunner.** 1985. On the nature of rabies virus-cellular receptor interactions, p. 77–81. *In* E. Kuwert, C. Merieux, H. Koprowski, and K. Bogel (ed.), Rabies in the tropics. Springer-Verlag, Vienna.
16. **Schlegel, R., T. S. Tralka, M. C. Willingham, and I. Pastan.** 1983. Inhibition of VSV binding and infectivity by phosphatidylserine: is phosphatidylserine a VSV binding site? Cell **32:**639–646.
17. **Schultz, M., and R. L. Crowell.** 1980. Acquisition of susceptibility to coxsackievirus A2 by the rat L8 cell line during myogenic differentiation. J. Gen. Virol. **46:**39–49.
18. **Seganti, L., M. Grassi, P. Mastromorino, A. Pana, F. Superti, and N. Orsi.** 1983. Activity of human serum lipoproteins on the infectivity of rhabdoviruses. Microbiologica **6:**91–99.
19. **Superti, F., M. Derer, and H. Tsiang.** 1984. Mechanism of rabies virus entry into CER cells. J. Gen. Virol. **65:**781–789.
20. **Superti, F., L. Seganti, H. Tsiang, and N. Orsi.** 1984. Role of phospholipids in rhabdovirus attachment to CER cells. Arch. Virol. **81:**321–328.
21. **Sytkowski, A. J., Z. Vogel, and M. W. Nirenberg.** 1973. Development of acetylcholine receptor clusters on cultured muscle cells. Proc. Natl. Acad. Sci. USA **70:**270–274.
22. **Tardieu, M., R. L. Epstein, and H. L. Weiner.** 1982. Interaction of viruses with cell surface receptors. Int. Rev. Cytol. **80:**27–61.
23. **Watson, J. D., G. H. Tignor, and A. L. Smith.** 1981. Entry of rabies virus into the peripheral nerves of mice. J. Gen. Virol. **56:**371–382.
24. **Wiktor, T. J., and H. F. Clark.** 1975. Growth of rabies virus in tissue culture, p. 155–179. *In* G. M. Baer (ed.), The natural history of rabies, vol. 1. Academic Press, Inc., New York.
25. **Wunner, W. H., K. J. Reagan, and H. Koprowski.** 1984. Characterization of saturable binding sites for rabies virus. J. Virol. **50:**691–697.
26. **Yaffe, D., and O. Saxel.** 1977. A myogenic cell line with altered serum requirements for differentiation. Differentiation **7:**159–166.

Biological Significance of the Epstein-Barr Virus Receptor on B Lymphocytes†

GLEN R. NEMEROW, MARTIN F. E. SIAW, AND NEIL R. COOPER

Department of Immunology, Scripps Clinic & Research Foundation, La Jolla, California 92037

Epstein-Barr virus (EBV), a human herpesvirus, selectively infects human B lymphocytes via a specific virus receptor which has been identified as the CR2 complement receptor. We have found that a B-cell-specific monoclonal antibody, OKB7, directly blocks EBV attachment to and infection of B cells. HB-5, another anti-CR2 monoclonal antibody, blocks these functions only in the presence of a second antibody; both monoclonal antibodies immunoprecipitate the same 145-kilodalton membrane protein. EBV is a T-independent B-cell mitogen which also induces polyclonal immunoglobulin synthesis. OKB7, HB-5, and another CR2 monoclonal antibody, anti-B2, were examined for ability to mimic EBV and induce B-cell activation. OKB7 and its F(ab')2 fragment stimulated DNA synthesis of unseparated human peripheral blood lymphocytes as measured by [³H]thymidine incorporation; anti-B2 and HB-5 had negligible mitogenic activity. B but not T cells were responsible for DNA synthesis. Unlike EBV-induced mitogenesis, however, T cells were required for B-cell activation by OKB7. The purified C3d,g complement fragment, which also reacts with CR2, failed to activate B cells whether tested in solution or attached to fluorescent microspheres. OKB7 also induced B-cell differentiation as documented by polyclonal immunoglobulin secretion (IgM IgG IgA); immunoglobulin synthesis was also T dependent. Thus the EBV/CR2 receptor possesses multiple antigenic and functional epitopes, some of which induce B-lymphocyte activation and differentiation.

Epstein-Barr virus (EBV), a human herpesvirus, causes or is associated with a number of diseases, including infectious mononucleosis (14), Burkitt's lymphoma (13), nasopharyngeal carcinoma (3), and several autoimmune diseases (34). The virus is oncogenic in subhuman primates (26); in vitro it transforms human B lymphocytes and thereby generates polyclonal lymphoblastoid cell lines (5). EBV exhibits a highly restricted cell tropism: in vitro it selectively infects B lymphocytes (11, 17, 30) and probably nasoepithelial cells (33), and only these cell types obtained from patients contain the viral genome. Infection of B lymphocytes is initiated by specific binding of EBV to a receptor expressed on the surface of B lymphocytes and B lymphoblastoid cell lines (17–19, 30). Several different lines of evidence have suggested that the viral receptor either is a C3 receptor or is closely associated with a C3 receptor. Thus,

†Publication no. 3887 IMM.

both receptors are coexpressed on many cell lines (5, 18, 21) and co-induced in receptor-negative cell lines (23). The two receptors co-cap (18, 40), and both are stripped from the membrane by the same treatments (39). Association is further suggested by the findings that C3 blocks EBV fusion with cellular membranes (32), rosetting of C3-coated erythrocytes is impaired by virus binding (15, 18), and sequential incubation with activated C3, anti-C3, and anti-immunoglobulin blocks EBV binding (18). These several properties correlate with expression of the CR2 or C3d receptor rather than the CR1 or C3b receptor (6, 16, 19). A recent study concluded that the CR2 and EBV receptors are different structures (15). This work used a 72-kilodalton (kDa) protein, isolated from the culture super- natant of Raji lymphoblastoid cells as "CR2" (22). However, CR2 has been clearly shown to be a 140- to 145-kDa protein by several laboratories (16, 38), and the 72-kDa protein may represent a degradation product of the native receptor. The 140- to 145-kDa protein is recognized by monoclonal antibodies termed anti-B2 (16, 28) and HB-5 (6, 37), both of which immunoprecipitate a polypeptide chain of this molecular mass from surface-labeled normal B lymphocytes and B lymphoblastoid cell lines. Both anti-B2 and HB-5, together with a second anti-immunoglobulin antibody, block rosette formation between particles bearing C3d and B lymphoid cells bearing the receptor (16, 38). Anti-B2 also has been found to have modest ability to directly block such rosettes in the absence of a second antibody (16). HB-5, although unable to directly block EBV binding, possessed this ability in the presence of a second anti-immunoglobulin antibody (6). Very recent studies with HB-5 have pro- vided additional evidence for the identity of the EBV receptor with CR2, since pretreatment of B lymphoblastoid cells with HB-5 followed by anti-immunoglo- bulin blocked EBV binding, and HB-5 CR2 complexes immobilized on *Staphylo- coccus aureus* particles possessed the ability to bind EBV (6). Polyclonal anti-CR2 has been found to block EBV binding in the absence of a second anti-immunoglo- bulin antibody (7). These data strongly support the contention that CR2 and EBV receptor functions are two properties of the same membrane protein.

Our recent studies have been concerned with the mechanism of entry of EBV into normal B lymphocytes (30). These studies indicate that EBV enters normal B lymphocytes by an endocytic pathway differing in several regards from the clathrin-receptosome-lysosome pathway utilized by a number of other ligands, including several viruses (10, 12, 24, 36); in contrast, EBV enters Raji lympho- blastoid cells by direct fusion with the external cell membrane (30). In initiating studies of the nature and normal biological function of the structure to which EBV binds, we screened eight B-lymphocyte-specific monoclonal antibodies for ability to directly block binding of radiolabeled purified EBV (4, 29) to isolated human tonsil B cells. Four of these monoclonal antibodies, HB-5, anti-B2, OKB7, and AB-1, immunprecipitated a 145-kDa B-cell membrane glycoprotein (6, 16, 28, 31, 37, 38; G. R. Nemerow, M. E. McNaughton, and N. R. Cooper, submitted for publication; B. S. Wilson, J. L. Platt, and N. E. Kay, submitted for publication), and cross-absorption studies indicated that the antibodies recog- nized the same membrane protein (31; Wilson et al., submitted). The other four monoclonal antibodies, anti-B1, anti-B4, anti-Leu 14, and OKB2, immuno- precipitated membrane antigens of different molecular weights. Of the eight monoclonal antibodies, only OKB7 directly blocked EBV binding to tonsil B lymphocytes (31; Nemerow et al., submitted). Results of such studies with OKB7 and anti-B2 are shown in Fig. 1. Inhibition of virus binding was not related to the antibody isotype, as HB-5, AB-1, and anti-B1, all of which, like OKB7, are of the immunoglobulin G2a (IgG2a) isotype, failed to block binding. Inhibition of virus binding did not correlate with the percentage of B cells recognized by the antibodies or with the staining intensity as determined by fluorescence with anti-B1, OKB7, and anti-B2 among several B cell lines as well as tonsil and peripheral blood B cells; T cells were unreactive (31). Inhibition of virus binding was dose dependent; 50% inhibition of binding of 10^7 particles of EBV to 10^7 tonsil B cells occurred with a dose of 300 ng of OKB7 (Fig. 1A).

Consistent with the inhibition of virus binding, OKB7 blocked EBV infection of peripheral B lymphocytes in a dose-dependent manner whether measured by colony outgrowth in soft agarose or by [^{3}H]thymidine incorporation after 14 days of culture (Fig. 1B); the other antibodies were negative. Fifty percent inhibition of infection of 6×10^6 peripheral blood B cells required 300 to 500 ng of OKB7.

EBV is a T-cell-independent B-cell mitogen and induces polyclonal immunoglobulin synthesis and secretion (2, 9, 20). We examined whether HB-5, anti-B2, OKB7, and AB-1 would mimic EBV-induced mitogenic stimulation. In this regard, OKB7 had earlier been reported to modulate B-cell activation by other stimuli (27). In addition, polyclonal anti-CR2 was very recently reported to produce a modest twofold enhancement of DNA synthesis in human peripheral blood mononuclear cells in the presence of T-cell factors (8). We found HB-5 to have negligible mitogenic activity (Nemerow et al., submitted), a finding which confirms studies of others (37), while anti-B2 had marginal activity. In marked contrast, OKB7 and AB-1 produced a dose-dependent enhancement of [^{3}H]-thymidine incorporation in unseparated peripheral blood B lymphocytes; at maximal doses, 50- to 100-fold increase was observed (Nemerow et al., submitted). Results of typical studies with anti-B2 and OKB7 are shown in Fig. 2. The F(ab')$_2$ fragment of OKB7 was as active as the uncleaved molecule, indicating that the Fc receptor is not involved in B-cell activation mediated by the anti-CR2 antibodies. Separation of B and T cells after 6 days of culture revealed that B cells were responsible for the DNA synthesis (Nemerow et al., submitted). Unlike EBV, however, OKB7 failed to induce DNA synthesis in isolated B cells; the presence of T cells was found to be necessary for the mitogenic activity.

OKB7 also induced B-cell differentiation, i.e., polyclonal immunoglobulin secretion. One microgram of OKB7 per ml induced the secretion of 1,800 μg of IgM, 420 μg of IgG, and 84 μg of IgA after 10 days of culture. Immunoglobulin synthesis was also T-cell dependent.

The purified C3d,g complement fragment, which also reacts with CR2, was also tested for ability to activate B cells. Although avid binding to B lymphocytes was observed, C3d,g failed to activate either B cells or unseparated peripheral blood lymphocytes whether tested in solution or attached to fluorescent microspheres.

These studies confirm and further substantiate the earlier indications that a single B-cell membrane protein functions as a dual receptor for EBV and C3d,g or C3d. The protein contains multiple antigenic and functional epitopes subserving different functions (Table 1). The area of the CR2 molecule recognized

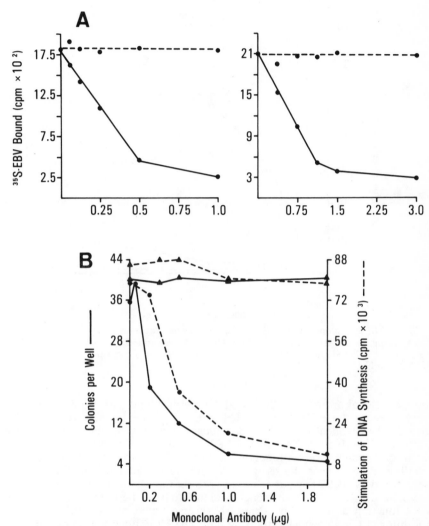

FIG. 1. Effect of monoclonal antibodies on ^{35}S-EBV binding (A) and infection (B). (A) Normal tonsil B cells (upper left) or Raji cells (upper right) (10^7) were allowed to react with various amounts of OKB7 (●———●) or anti-B2 (●– – –●) prior to addition of 4,000 cpm of ^{35}S-EBV. (B) B cells (6 × 10^5) were allowed to react with various amounts of OKB7 (●) or anti-B2 (▲), washed, and then cultured for 14 days in the presence of EBV. Infectivity was measured by [^{3}H]thymidine incorporation (– – – –) and by colony forma-tion (———).

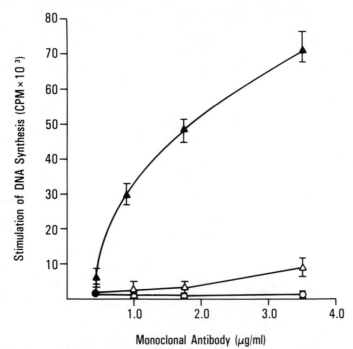

FIG. 2. Dose response of anti-EBV/CR2 monoclonal antibody-induced stimulation of DNA synthesis. Anti-B1 (O), anti-B2 (△), or OKB7 (▲) was added at various concentrations to peripheral blood mononuclear cells (10^5 per well) for 6 days prior to addition of [^{3}H]thymidine.

by monoclonal antibodies OKB7 and AB-1 either functions as or is sterically very close to the EBV receptor. This portion of the 145-kDa receptor also triggers T-dependent (with the monoclonal antibodies) or -independent (with EBV) B-cell activation. This area of CR2 overlaps the determinants concerned with complement receptor function, as OKB7 efficiently inhibits C3d,g and C3d binding in the absence of a second anti-immunoglobulin antibody. The epitope recognized by OKB7 appears to be closer to the C3d,g receptor than the epitope recognized by anti-B2 since monoclonal antibody to the latter did not inhibit CR2 function as efficiently. Others have also found anti-B2 to be a weak inhibitor of C3d-dependent rosetting (16). The B2 epitope is also not directly overlapping with the structures involved in EBV binding, nor is it directly overlapping with the HB-5 antigen (37). The epitope recognized by HB-5 is not mitogenic; HB-5 does not directly block EBV binding or C3d-dependent rosetting although both properties are inhibited in the presence of a second antibody (6, 38).

These studies thus indicate that EBV utilizes a normal cell constituent to achieve specific B-cell tropism. Certain features of the EBV receptor on B

TABLE 1. Functional properties of anti-EBV/CR2 receptor monoclonal antibodies

Monoclonal antibody	Inhibition of EBV binding[a] (1° Ab/2nd Ab)	Inhibition of C3d binding[b] (1° Ab/2nd Ab)	Proliferation	Differentiation
Anti-B2	−/−	±/±	±	−
AB-1	−/±	±/+	+	+
OKB7	+/+	+/+	+	+
HB-5	−/+	−/+	−	−

[a] Inhibition of EBV binding was determined by reaction of Raji cells with 1 to 2 µg of each monoclonal antibody (1° Ab) or with this antibody and a second immunoglobulin at 10 µg/ml (2nd Ab). Inhibition of ^{35}S-EBV binding was considered to be positive if greater than 50% blocking of virus binding was achieved.

[b] Inhibition of C3d binding was by effect on rosette formation of EAC3d cells with tonsil B cells or by fluorescent microsphere coated with C3d,g binding to Raji cells by the primary monoclonal antibody (1° Ab) or by this antibody and a second treatment with anti-mouse immunoglobulin (2nd Ab).

lymphocytes suggest further complexity. For example, CR2 is a differentiation marker for B lymphocytes, being found only on cells at a restricted stage of the B-cell maturation sequence (1, 28, 37). Alterations in the levels of the molecule with activation have been reported; thus, B cells activated with mitogen undergo an initial increase, followed by a decrease in the expression of the 145-kDa protein recognized by anti-B2 (35).

As shown here, only certain epitopes of the EBV/CR2 receptor are involved in cell activation. This type of selectivity, the mechanisms of which are unclear, has also been observed in T-cell activation after reaction with some, but not other, antibodies to the sheep erythrocyte receptor (25). The finding that certain epitopes of the EBV/CR2 receptor trigger B-lymphocyte proliferation and differentiation also raises the possibility that B-cell activation induced by EBV binding to the receptor facilitates, or is essential for, B-cell transformation by the virus. Such events could also play a role in the establishment of latent viral infection by this herpesvirus.

These various B-cell-specific monoclonal antibodies and other CR2 ligands, such as EBV, C3d,g, and C3d, which all react with the same protein on B lymphocytes, but induce very different biological functions, will provide useful probes not only for dissecting the early stages of EBV infection. Equally important, they will be invaluable for determining the role of C3 fragments in immune responses and for analyzing the biochemical events involved in a novel pathway of B-cell activation and differentiation.

We thank Bonnie Weier for preparing the manuscript.

This work was supported by Public Health Service grants CA35048, CA14692, and AI17354 from the National Institutes of Health and by a Leukemia Society of America Special Fellowship Award.

LITERATURE CITED

1. **Bhan, A. K., L. M. Nadler, P. Stashenko, R. T. McCluskey, and S. F. Schlossman.** 1981. Stages of B cell differentiation in human lymphoid tissue. J. Exp. Med. **154:**737–749.

166   CELLULAR RECEPTORS

2. **Bird, G., and S. Britton.** 1979. A new approach to the study of human B lymphocyte function using an indirect plaque assay and a direct B cell activator. Immunol. Rev. **45:**41.
3. **De-The, G., and Y. Zeng.** 1982. Epidemiology of Epstein-Barr virus: recent results on endemic virus non-endemic Burkitt's lymphoma and EBV related pre-nasopharyngeal carcinoma conditions, p. 3–15. *In* D. S. Yohn and J. R. Blakeslee (ed.), Advances in comparative leukemia research. Elsevier Biomedical Press, New York.
4. **Edson, C. M., and D. A. Thorley-Lawson.** 1981. Epstein-Barr virus membrane antigens: characterization, distribution, and strain differences. J. Virol. **39:**172–183.
5. **Einhorn, L., M. Steinitz, E. Yefenof, I. Ernberg, T. Bakacs, and G. Klein.** 1978. Epstein-Barr virus (EBV) receptors, complement receptors, and EBV infectibility of different lymphocyte fractions of human peripheral blood. II. Epstein-Barr virus studies. Cell. Immunol. **35:**43–58.
6. **Fingeroth, J. D., J. J. Weis, T. F. Tedder, J. L. Strominger, P. A. Biro, and D. T. Fearon.** 1984. Epstein-Barr virus receptor of human B lymphocytes is the C3d receptor CR2. Proc. Natl. Acad. Sci. USA **81:**4510–4514.
7. **Frade, R., M. Barel, B. Ehlin-Henriksson, and G. Klein.** 1985. gp140, the C3d receptor of human B lymphocytes, is also the Epstein-Barr virus receptor. Proc. Natl. Acad. Sci. USA **82:**1490–1493.
8. **Frade, R., M. Crevon, M. Barel, A. Vazquez, L. Krikorian, C. Charriant, and P. Galenaud.** 1985. Enhancement of human B cell proliferation by an antibody to the C3d receptor, the gp 140 molecule. Eur. J. Immunol. **15:**73–76.
9. **Gerber, P., and B. H. Hoyer.** 1972. Induction of cellular DNA synthesis in human leukocytes by Epstein-Barr virus. Nature (London) **231:**46–47.
10. **Goldstein, J. L., R. G. W. Anderson, and M. S. Brown.** 1979. Coated pits, coated vesicles, and receptor-mediated endocytosis. Nature (London) **279:**679–685.
11. **Greaves, M. F., G. Brown, and A. B. Rickinson.** 1975. Epstein-Barr virus binding sites on lymphocyte subpopulations and the origin of lymphoblasts in cultured lymphoid cell lines and in the blood of patients with infectious mononucleosis. Clin. Immunol. Immunopathol. **3:**514–524.
12. **Helenius, A., J. Kartenbeck, K. Simons, and E. Fries.** 1980. On the entry of Semliki Forest virus into BHK-21 cells. J. Cell. Biol. **84:**404–420.
13. **Henle, W., V. Diehl, G. Kohn, H. Zur-Hausen, and G. Henle.** 1967. Herpes-type virus and chromosome marker in normal leukocytes after growth with irradiated Burkitt cells. Science **157:**1064–1065.
14. **Henle, G., W. Henle, and V. Diehl.** 1968. Relationship of Burkitt tumor associated herpes-type virus to infectious mononucleosis. Proc. Natl. Acad. Sci. USA **59:**94–101.
15. **Hutt-Fletcher, L. M., E. Fowler, J. D. Lambris, R. J. Feighny, J. G. Simmons, and G. D. Ross.** 1983. Studies of the Epstein-Barr virus receptor found on Raji cells. II. A comparison of lymphocyte binding sites for Epstein-Barr virus and C3d. J. Immunol. **130:**1309–1312.
16. **Iida, K., L. Nadler, and V. Nussenzweig.** 1983. Identification of the membrane receptor for the complement fragment C3d by means of a monoclonal antibody. J. Exp. Med. **158:**1021–1033.
17. **Jondal, M., and G. Klein.** 1973. Surface markers on human B and T lymphocytes. II. Presence of Epstein-Barr receptors on B lymphocytes. J. Exp. Med. **138:**1365–1378.
18. **Jondal, M., G. Klein, M. B. A. Oldstone, V. Bokisch, and E. Yefenof.** 1976. Surface markers on human B and T lymphocytes. VIII. Association between complement and Epstein-Barr receptors on human lymphoid cells. Scand. J. Immunol. **5:**401–410.
19. **Jonsson, V., A. Wells, and G. Klein.** 1982. Receptors for the complement C3d component and the Epstein-Barr virus are quantitatively co-expressed on a series of B cell lines and their derived somatic cell hybrids. Cell. Immunol. **72:**265–276.
20. **Kirchner, H., G. Tosato, M. R. Blaese, S. Broder, and I. Magrath.** 1979. Polyclonal immunoglobulin secretion by human B lymphocyte exposed to Epstein-Barr virus in vitro. J. Immunol. **122:**1310–1313.
21. **Klein, G., E. Yefenof, K. Falk, and A. Westman.** 1978. Relationship between Epstein-Barr virus (EBV) production and the loss of the EBV receptor/complement receptor complex in a series of sublines derived from the same original Burkitt's lymphoma. Int. J. Cancer **21:**552–560.
22. **Lambris, J. D., N. J. Dobson, and G. D. Ross.** 1981. Isolation of lymphocyte membrane complement receptor type two (the C3d receptor) and preparation of receptor-specific antibody. Proc. Natl. Acad. Sci. USA **78:**1828–1832.
23. **Magrath, I., C. Freeman, M. Santaella, J. Gadek, M. Frank, R. Spiegel, and L. Novikovs.** 1981. Induction of complement receptor expression in cell lines derived from human undifferentiated lymphomas. II. Characterization of the induced complement receptors and demonstration of the simultaneous induction of EBV receptor. J. Immunol. **127:**1039–1043.

24. **Matlin, K., H. Reggio, A. Helenius, and K. Simons.** 1981. Infectious entry pathway of influenza virus in a canine kidney cell line. J. Cell Biol. **91**:601–603.

25. **Meuer, S. C., R. E. Hussey, M. Fabbi, D. Fox, O. Acuto, K. A. Fitzgerald, J. C. Hodgdon, J. P. Protentes, S. F. Schlossman, and L. Reinherz.** 1984. An alternative pathway of T-cell activation: a functional role for the 50 Kd T11 sheep erythrocyte receptor protein. Cell **36**:897.

26. **Miller, G., T. Shope, H. Lisco, D. Stitt, and M. Lipman.** 1972. Epstein-Barr virus: transformation, cytopathic changes, and viral antigens in squirrel monkey and marmoset leukocytes. Proc. Natl. Acad. Sci. USA **69**:383–387.

27. **Mittler, R. S., M. A. Talle, K. Carpenter, P. E. Rao, and G. Goldstein.** 1983. Generation and characterization of monoclonal antibodies reactive with human B lymphocytes. J. Immunol. **131**:1754–1761.

28. **Nadler, L. M., P. Stashenko, R. Hardy, A. van Agthoven, C. Terhorst, and S. F. Schlossman.** 1981. Characterization of a human B cell specific antigen (B2) distinct from B1. J. Immunol. **126**:1–41–1947b.

29. **Nemerow, G. R., and N. R. Cooper.** 1981. Isolation of Epstein-Barr virus and studies of its neutralization by human IgG and complement. J. Immunol. **127**:272–278.

30. **Nemerow, G. R., and N. R. Cooper.** 1984. Early events in the infection of human B lymphocytes by Epstein-Barr virus: the internalization process. Virology **132**:186–198.

31. **Nemerow, G. R., R. Wolfert, M. E. McNaughton, and N. R. Cooper.** 1985. Identification and characterization of the Epstein-Barr virus receptor on human B lymphocytes and its relationship to the C3d complement receptor (CR2). J. Virol. **55**:347–351.

32. **Rosenthal, K. S., S. Yanovich, M. Inbar, and J. L. Strominger.** 1978. Translocation of a hydrocarbon fluorescent probe between Epstein-Barr virus and lymphoid cells: an assay for early events in viral infection. Proc. Natl. Acad. Sci. USA **75**:5076–5080.

33. **Sixbey, J. W., E. H. Vesterinen, J. G. Nedrud, N. Raab-Traub, L. A. Walton, and J. S. Pagano.** 1983. Replication of Epstein-Barr virus in human epithelial cells infected *in vitro*. Nature (London) **306**:480–483.

34. **Slaughter, L., D. A. Carson, F. C. Jensen, T. L. Holbrook, and J. H. Vaughn.** 1978. In vitro effects of Epstein-Barr virus on peripheral blood mononuclear cells from patients with rheumatoid arthritis and normal subjects. J. Exp. Med. **148**:1429–1434.

35. **Stashenko, P., L. M. Nadler, R. Hardy, and S. F. Schlossman.** 1981. Expression of cell surface markers after human B lymphocyte activation. Proc. Natl. Acad. Sci. USA **78**:3848–3852.

36. **Steinman, R. M., I. S. Mellman, W. A. Muller, and Z. A. Cohn.** 1983. Endocytosis and the recycling of plasma membrane. J. Cell Biol. **96**:1–27.

37. **Tedder, T. F., L. T. Clement, and M. D. Cooper.** 1984. Expression of C3d receptors during human B cell differentiation: immunofluorescence analysis with the HB-5 monoclonal antibody. J. Immunol. **133**:678–683.

38. **Weis, J. J., T. F. Tedder, and D. T. Fearon.** 1984. Identification of a 145,000 Mr membrane protein as the C3d receptor (CR2) of human B lymphocytes. Proc. Natl. Acad. Sci. USA **81**:881–885.

39. **Yefenof, E., and G. Klein.** 1977. Membrane receptor stripping confirms the association between EBV receptors and complement receptors on the surface of human B lymphoma lines. Int. J. Cancer **20**:347–352.

40. **Yefenof, E., G. Klein, M. Jondal, and M. B. A. Oldstone.** 1976. Surface markers on human B and T lymphocytes. IX. Two color immunofluorescence studies on the association between EBV receptors and complement receptors on the surface of lymphoid cell lines. Int. J. Cancer **17**:693–700.

Penetration and Uncoating

Entry Mechanisms of Picornaviruses

SJUR OLSNES, INGER HELENE MADSHUS, AND KIRSTEN
SANDVIG

*Norsk Hydro's Institute for Cancer Research and The Norwegian Cancer Society,
Montebello, Oslo, Norway*

**The entry into cells of three picornaviruses, poliovirus, human
rhinovirus 2, and encephalomyocarditis virus, has been studied with
virus rendered light sensitive by growth in the presence of neutral
red and acridine orange. The entry of poliovirus and human
rhinovirus 2 was strongly inhibited by monensin, protonophores,
and tributyl-tin, which increase the pH of intracellular vesicles,
indicating that the transfer of the virus genome across the cellular
membrane occurs in acidified vesicles rather than at the cell surface.
In the case of poliovirus, but not with human rhinovirus 2, the
protection could be overcome when cells with surface-bound virus
were exposed to medium with low pH. Experiments with the
detergent Triton X-114 showed that poliovirus entered the detergent
phase at low pH, but not at neutral pH, indicating that at the low
pH normally hidden hydrophobic domains of the capsid proteins
become exposed. Possibly these domains induce association of the
capsid with the membrane and release of the genome at the
cytosolic side. Encephalomyocarditis virus does not appear to require low pH for its penetration into the cytosol. The entry mechanism of the picornaviruses is compared with that of protein toxins
with intracellular sites of action.**

During recent years our knowledge on the mechanisms of entry of enveloped
viruses into eucaryotic cells has increased considerably. The ability of Sendai
virus to fuse with cellular membranes has been known for several years (4), but
the generality of the principle first became evident with the observation by
Helenius et al. (5, 12) that a large number of enveloped viruses that do not fuse
with membranes at neutral pH do so if the pH is lowered to 5 to 6. This
observation focused attention on intracellular acidic vesicles as the actual site
of transfer of the virus genome into the cytosol. The mechanism of the fusion of
the virus envelope with cell membranes has been discussed in considerable
detail (6). However, a large number of important viruses do not contain a lipid
envelope that can fuse with cellular membranes and thereby facilitate entry of
the virus genome.

In our laboratory we have for several years studied the entry into the cytosol
of a number of protein toxins such as diphtheria toxin, ricin, abrin, and others
(14; S. Olsnes, K. Sandvig, I. H. Madshus, and A. Sundan, Biochem. Soc. Symp.,
in press). These toxins consist of two disulfide-linked polypeptides, one of which
binds the toxins to cell surface receptors while the other enters the cytosol and
inactivates in an enzymatic fashion vital components of the protein synthesis

machinery (13, 15). A single A-chain molecule in the cytosol is sufficient to kill the cell. It occurred to us that at least certain noncoated viruses like the picornaviruses may use routes similar to those of the toxins in entering the cytosol. The picornaviruses are positive-strand RNA viruses, and the only requirement for infection is the successful transfer of one RNA molecule into the cytosol. Therefore, in both cases the transfer of a hydrophilic macromolecule across the membrane is the only requirement for biological effect.

Among the protein toxins, diphtheria toxin is best characterized with respect to its entry mechanism. The B fragment of this toxin contains a hydrophobic region which is normally hidden (1). When the toxin is exposed to pH below 5.5, this region becomes evident and may then insert itself into lipid membranes (17). Under normal conditions the exposure of receptor-bound diphtheria toxin to low pH in endosomes leads to translocation of the A fragment across the membrane. However, if cells with surface-bound toxin are exposed to acidic medium, entry may be induced from the cell surface (16).

In uptake studies with bacterial and plant toxins, we have extensively used various inhibitors. Even inhibitors that are toxic to the cells after prolonged exposure can be used since the biological effect of the toxins develops rapidly. In the case of picornaviruses, the easiest way to monitor virus entry is to measure virus production. Since this process requires several hours to develop, it was necessary to devise a test system in which the exposure of the cells to the inhibitors could be limited to the time period of actual virus entry.

LIGHT-SENSITIVE PICORNAVIRUSES

It was observed already in the beginning of this century that a number of viruses are sensitized to visible light by the presence of various acridine dyes. Apparently, upon exposure to light, the dyes form reactive compounds which inactivate the virus genome. Mature picornaviruses incubated with acridine dyes do not become light sensitive, because their capsids are impermeable to the dyes. Picornaviruses may, however, be rendered light sensitive if they are produced in cells growing in the presence of the dyes (3, 11, 19). Under these conditions the dye is trapped inside the virus.

When light-sensitive picornaviruses are incubated with virus-sensitive cells, the RNA remains light sensitive only as long as the RNA is contained inside the capsid (11). The transition of the virus from light sensitivity to light resistance can therefore be taken as a measure of penetration of the viral genome into the cytosol. It is not known at what stage in the infection process the virus becomes light resistant and whether this is a sudden event. In work with light-sensitive virus, it is therefore possible to draw conclusions only on the part of the infection process that occurs before the virus becomes light resistant. This may not include the whole entry process.

Enteroviruses can easily be rendered light sensitive by growing them in the presence of neutral red (8, 11). However, we found acridine orange preferable for production of photosensitive human rhinovirus 2 (HRV2) (19). The reason for this apparently is that neutral red, being a weak base, penetrates the cells, increases the pH of the endocytic vesicles, and inhibits HRV2 infection (see below).

Since the entry period of picornaviruses is comparatively short, the use of light-sensitive viruses allowed us to study the effect on the uptake process of a variety of inhibitors that the cells would not tolerate for prolonged periods of time.

REQUIREMENT OF LOW pH FOR VIRUS ENTRY

The ability of HRV2 to infect cells was strongly inhibited in the presence of NH_4Cl and a number of other weak bases (9). The cells were also protected by the ionophore monensin, which exchanges monovalent cations for protons across membranes and thus dissipates proton gradients. Also tributyl-tin, which dissipates proton gradients by hydroxyl transfer, and protonophores like FCCP (carbonylcyanide p-trifluoro-methoxy-phenylhydrazone) protected against HRV2.

Radioactively labeled HRV2 binds comparatively slowly to HeLa cells. The reason for this could be that a small number of receptors recycle rapidly and that the apparent binding in fact represents uptake by receptor-mediated endocytosis (9). The possibility therefore existed that the protective effect of compounds which increase intravesicular pH is due to inhibited recycling of endocytosed HRV2 receptors and, as a consequence, reduced ability of the cells to accumulate the virus. Experiments with [^{35}S]methionine-labeled virus showed that the protective compounds indeed have such an effect (Fig. 1).

To test whether the reduction in virus binding was the only reason for protection, we repeated the experiment under conditions where we first allowed the virus to bind at 0°C to untreated cells and then washed the cells to remove unbound virus. When the cells were subsequently incubated at 37°C in the presence of the inhibitors, they were protected against the virus (Fig. 2). This indicates that low pH in intracellular vesicles is necessary for the entry of bound HRV2 as well as for the recycling of endocytosed virus receptors back to the cell surface.

Also, poliovirus was found to require low pH for entry. Thus, treatment of the cells with procaine, monensin, protonophores, and tributyl-tin all strongly protected the cells against poliovirus infection (Fig. 3). In the case of poliovirus there was no evidence for reduced binding after treatment with the different inhibitors.

The ability of NH_4Cl and some other weak bases to protect against poliovirus was lower than expected because they have a dual effect on poliovirus infection (8). Thus, while the amines were found to inhibit the entry process, they stimulated virus production in those cells that were infected, perhaps as a result of a slight increase in the cytoplasmic pH. Therefore, in a test system based on measurements of virus production, the protection was less than the inhibitory effect on virus entry as such.

When poliovirus was added to cells in the presence of monensin, which inhibits infection by the normal route, the protection could be overcome by exposing the cells to low pH (Fig. 4). It therefore appears that, in the case of poliovirus, low pH can induce infection, possibly directly through the surface membrane. Half-maximal infection occurred at approximately pH 6.1 (10).

Compounds that increase the pH of intracellular vesicles, such as procaine, NH_4Cl, monensin, and FCCP, had little or no effect on the entry of enceph-

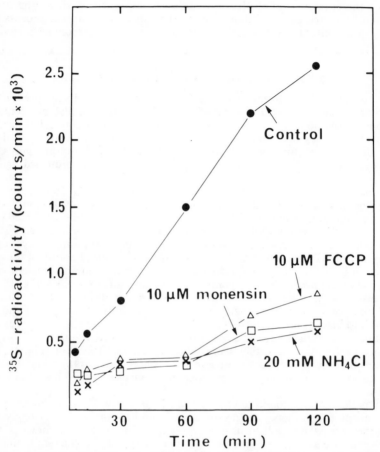

FIG. 1. Effect of various compounds on the binding of HRV2 to cells. Cells were preincubated with the compounds indicated for 15 min at 37°C, and then the ability of the cells to bind [^{35}S]methionine-labeled virus at 18°C was measured. (From reference 9.)

alomyocarditis virus into cells. Furthermore, low pH in the medium inhibited rather than enhanced the entry of receptor-bound encephalomyocarditis virus. It therefore appears that low pH is not required for entry of this virus.

EFFECT OF LOW pH ON THE POLIOVIRUS PARTICLE

When poliovirus is taken up by cells, a considerable fraction of the virus is transformed into A particles, which have lost the capsid protein VP4 and which are no longer infective, although they retain the RNA genome. These particles

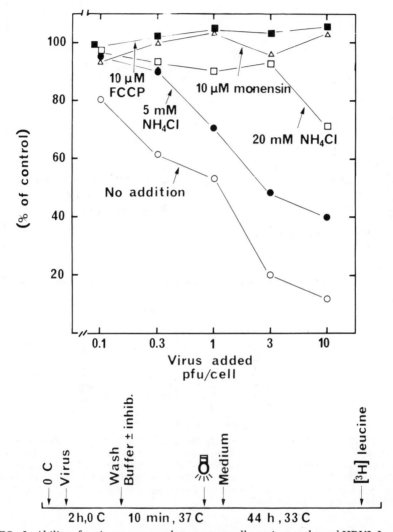

FIG. 2. Ability of various compounds to protect cells against prebound HRV2. Increasing amounts of light-sensitive HRV2 were added to cells at 0°C in the dark and allowed to bind for 2 h. Then the cells were washed and incubated at 37°C in medium with the compounds indicated. After 10 min the cells were exposed to visible light and transferred to normal medium. The cells were incubated at 33°C for 44 h, and then their ability to incorporate [³H]leucine was measured. (From reference 9.)

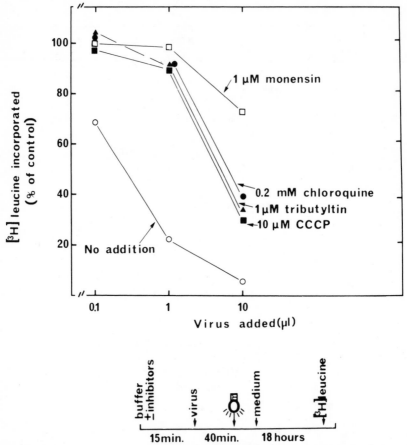

FIG. 3. Effect of different compounds on the ability of poliovirus to infect cells. Cells were incubated with the indicated compounds for 15 min at 37°C, and then increasing amounts of light-sensitive virus were added. After 40 min the cells were exposed to light, transferred to normal medium, and incubated for 18 h. Finally, the ability of the cells to incorporate [³H]leucine was measured. (Redrawn after reference 8.) CCCP, Carbonyl cyanide *m*-chlorophenylhydrazone.

are hydrophobic (7) and may represent an intermediary stage in the infection process.

It was found with all conditions tested that an increase in the pH of intracellular vesicles inhibited the transformation of the virus to A particles (8). Furthermore, when the protection by monensin was overcome by exposing the cells to low pH, A particles were formed. This indicates that the formation of A particles is closely connected with the infection process, either as a necessary

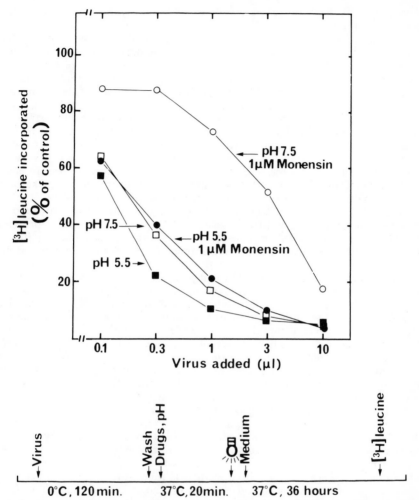

FIG. 4. Low-pH-induced virus entry into monensin-treated cells. Increasing amounts of light-sensitive poliovirus were added to cells and allowed to bind for 2 h at 0°C. Then the cells were washed, and medium with and without monensin adjusted to pH 5.5 or 7.5 was added. After 20 min at 37°C the cells were illuminated, changed to normal medium, and incubated for 36 h. Finally, the ability of the cells to incorporate [³H]leucine was measured. (Redrawn after reference 8.)

intermediate step or as a by-product formed by virus particles that are not successful in entering the cell (10).

The infection, as well as the formation of A particles, required physiological temperature to occur. Thus, both processes occurred very slowly at 27°C, whereas at 33°C both processes were clearly demonstrable and at 37°C they occurred rapidly (10).

The formation of A particles is not necessarily connected with virus entry. Thus, when the cytosol was acidified with acetic acid, poliovirus did not infect the cells although the treatment did not inhibit the formation of A particles. Apparently, for entry to occur it is not sufficient that virus present on the cell surface is altered by the low pH treatment. A pH gradient across the membrane also appears to be required for entry.

As with diphtheria toxin, the hemagglutinin glycoprotein of influenza virus contains a hidden hydrophobic domain which becomes exposed at low pH. This facilitates the fusion process between the virus envelope and the cellular membrane (18). To study the possibility that a similar process occurs with poliovirus, we measured the ability of [^{35}S]methionine-labeled virus to enter the detergent Triton X-114. This detergent forms supramicellar complexes and precipitates out of a water solution at temperatures above 20°C, while at 0°C it is dissolved as micelles. It has been shown that a number of hydrophobic proteins enter the detergent-rich phase at 37°C while hydrophilic proteins remain in the detergent-poor phase (2). Diphtheria toxin remained in the detergent-poor phase at neutral pH and entered the detergent-rich phase at pH 4.5.

Also, poliovirus stayed in the detergent-poor phase at neutral pH, but when the pH was reduced below 5, an increasing fraction of the virus entered the detergent-rich phase (Fig. 5). This indicates that at low pH the virus exposed normally hidden hydrophobic domains. Binding of the virus to cell surface receptors appears to facilitate the exposure of the hydrophobic domains. Thus, considerably less reduction in pH was required for entry of cell-bound virus into the detergent-rich phase than when the experiment was carried out with free virus. This cannot be due to altered properties of the detergent caused by the presence of lipids extracted from the cells. Thus, when the detergent was first treated with cells and then virus was added, slightly lower pH was required for entry of the virus into the detergent-rich phase than when untreated detergent was used. When poliovirus was exposed to pH 4.5 in the absence of cells, there was no formation of A particles at physiological temperature. However, when the virus was previously bound to paraformaldehyde-fixed cells, A particles were formed in high yield. This indicates that the exposure of hydrophobic domains which takes place in the free virus is reversible and that it becomes irreversible only after contact of the virus with the cell membrane.

CURRENT VIEW ON THE ENTRY MECHANISM

From the fact that the three different picornaviruses examined show different pH requirements for entry, it is clear that the entry mechanisms for the three viruses must be partly or entirely different. In the case of poliovirus and HRV2, which require low pH for entry, it is likely that the entry occurs from acidic intracellular vesicles. Taken with the fact that all three viruses enter the cytosol rapidly once they are bound to the cell surface, it is possible that entry occurs from newly formed vesicles, such as early endosomes. Once poliovirus bound to its receptor enters an acidic compartment, it is likely to expose hydrophobic

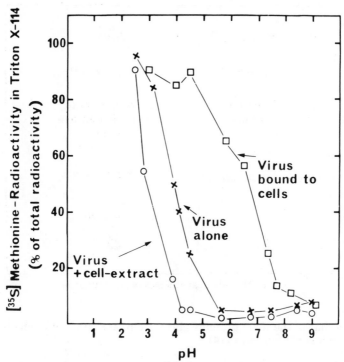

FIG. 5. Effect of pH on the ability of poliovirus to bind Triton X-114. [³⁵S]methionine-labeled virus was mixed with a solution of Triton X-114 and kept at 4°C for 15 min. Then the mixture was heated to 37°C to induce phase separation, and the sample was centrifuged for 2 min. The amount of radioactivity in the detergent-rich and detergent-poor phases was measured. The experiment was carried out with virus alone (×), with virus prebound to formaldehyde-fixed cells and then extracted with the detergent (□), and with virus added to a sample to Triton X-114 which had previously been treated with formaldehyde-fixed cells (○). (From reference 10.)

domains of the capsid proteins which may then insert themselves into the cellular membrane. This insertion could translocate the RNA genome to the cytoplasmic side of the membrane. The A particles formed concomitantly may represent virus particles that, as a result of inappropriate contact with the membrane upon acidification, were unable to insert themselves into the lipid bilayer (Fig. 6).

Encephalomyocarditis virus does not appear to require low pH for entry. Thus, compounds that increase the pH of intracellular vesicles did not protect against the virus, and low-pH treatment of cells with prebound virus did not induce entry of the virus but rather protected the cells against infection (9). This is in accordance with the finding that encephalomyocarditis virus in solution is inactivated at low pH. To some extent the uptake of encephalomyocarditis virus

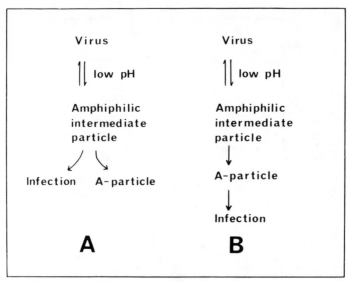

FIG. 6. Hypothetical schemes of poliovirus entry.

resembles that of the toxic proteins abrin, ricin, and viscumin, which also do not require low pH for entry (14).

LITERATURE CITED

1. **Boquet, P., M. S. Silverman, A. M. Pappenheimer, Jr., and W. B. Vernon.** 1976. Binding of Triton X-100 to diphtheria toxin, crossreacting material 45, and their fragments. Proc. Natl. Acad. Sci. USA **73:**4449–4453.
2. **Bordier, C.** 1981. Phase separation of integral membrane proteins in Triton X-114 solution. J. Biol. Chem. **256:**1604–1607.
3. **Crowther, D., and J. L. Melnick.** 1961. The incorporation of neutral red and acridine orange into developing poliovirus particles making them photosensitive. Virology **14:**11–21.
4. **Fan, D. P., and B. M. Sefton.** 1978. The entry into host cells of Sindbis virus, vesicular stomatitis virus and Sendai virus. Cell **15:**985–992.
5. **Helenius, A., J. Kartenbeck, K. Simons, and E. Fries.** 1980. On the entry of Semliki Forest virus into BHK-21 cells. J. Cell Biol. **84:**404–420.
6. **Lenard, J., and D. K. Miller.** 1983. Entry of enveloped viruses into cells, p. 119–138. *In* P. Cuatrecasas and T. F. Roth (ed.), Receptors and recognition, series B, vol. 15, Receptor-mediated endocytosis. Chapman & Hall, Ltd., London.
7. **Lonberg-Holm, K., L. B. Gosser, and E. J. Shimshick.** 1976. Interaction of liposomes with subviral particles of poliovirus type 2 and rhinovirus type 2. J. Virol. **19:**746–749.
8. **Madshus, I. H., S. Olsnes, and K. Sandvig.** 1984. Mechanism of entry into the cytosol of poliovirus type 1: requirement for low pH. J. Cell Biol. **98:**1194–1200.
9. **Madshus, I. H., S. Olsnes, and K. Sandvig.** 1984. Different pH requirements for entry of the two picornaviruses, human rhinovirus 2 and murine encephalomyocarditis virus. Virology **139:** 346–357.
10. **Madshus, I. H., S. Olsnes, and K. Sandvig.** 1984. Requirements for entry of poliovirus RNA into cells at low pH. EMBO J. **3:**1945–1950.
11. **Mandel, B.** 1967. The relationship between penetration and uncoating of poliovirus in HeLa cells. Virology **31:**702–712.

12. **Matlin, K. S., H. Reggio, A. Helenius, and K. Simons.** 1982. Pathway of vesicular stomatitis virus entry leading to infection. J. Mol. Biol. **156**:609–613.
13. **Olsnes, S., and A. Pihl.** 1982. Toxic lectins and related proteins, p. 51–105. *In* P. Cohen and S. van Heyningen (ed.), Molecular action of toxins and viruses. Elsevier Biomedical Press, New York.
14. **Olsnes, S., and K. Sandvig.** 1983. Entry of toxic proteins into cells, p. 187–236. *In* P. Cuatrecasas and T. F. Roth (ed.), Receptors and recognition, series B, vol. 15, Receptor-mediated endocytosis. Chapman & Hall, Ltd., London.
15. **Pappenheimer, A. M., Jr.** 1977. Diphtheria toxin. Annu. Rev. Biochem. **46**:69–94.
16. **Sandvig, K., and S. Olsnes.** 1980. Diphtheria toxin entry into cells is facilitated by low pH. J. Cell Biol. **87**:828–832.
17. **Sandvig, K., and S. Olsnes.** 1981. Rapid entry of nicked diphtheria toxin into cells at low pH. Characterization of the entry process and effects of low pH on the toxin molecule. J. Biol. Chem. **256**:9068–9076.
18. **Skehel, J. J., P. M. Bayley, E. B. Brown, S. R. Martin, M. D. Waterfield, J. M. White, I. A. Wilson, and D. C. Wiley.** 1982. Changes in the conformation of influenza virus hemagglutinin at the pH optimum of virus-mediated membrane fusion. Proc. Natl. Acad. Sci. USA **79**:968–972.
19. **Wilson, J. N., and P. D. Cooper.** 1963. Aspects of the growth of poliovirus as revealed by the photodynamic effects of neutral red and acridine orange. Virology **21**:135–145.

Uncoating of Vesicular Stomatitis Virus

JOHN LENARD

Department of Physiology and Biophysics, University of Medicine and Dentistry of New Jersey-Rutgers Medical School, Piscataway, New Jersey 08854

The sequence of events by which vesicular stomatitis virus becomes uncoated to initiate infection resembles that for other enveloped viruses and consists of the following steps. (i) The virus binds to suitable regions or molecules of the cell surface; these migrate to coated pit regions by normal cellular processes, carrying the virus with them. (ii) The viruses are internalized by endocytosis and subsequently appear in endosomes lacking the clathrin coat. (iii) The low pH (around 5) of the endosomes activates the fusogenic activity of the G protein on the viral surface, which catalyzes fusion between the viral envelope and the endosomal membrane, thus transferring the nucleocapsid to the cytoplasmic side. (iv) Virions that do not fuse in the endosomes may be passed to the lysosomes, where any further acid-mediating fusion must compete with degradation by the lysosomal hydrolases.

The sequence of events by which vesicular stomatitis virus (VSV) is currently thought to become uncoated to initiate infection in a cell is shown in Fig. 1. The virus binds to suitable regions or molecules on the cell surface; these migrate to coated pit regions by normal cellular processes, carrying the virus with them (Fig. 1, A). After internalization by endocytosis (Fig. 1, B), the viruses are found in endosomes lacking the clathrin coat (Fig. 1, C). The low pH (around 5) of the endosome activates the fusogenic activity of the G protein on the viral surface. This protein catalyzes fusion between the viral envelope and the endosomal membrane, thus transferring the viral nucleocapsid (containing the VSV genome RNA) to the cytoplasmic side of the endosomal membrane (Fig. 1, D). Virions that do not fuse in the endosomes may be passed to the lysosomes, where any further acid-mediated fusion must compete with degradation by the lysosomal hydrolases (Fig. 1, E). These events are virtually identical to those described for the entry of Semliki Forest virus and influenza virus into cells and have been suggested for the entry of other viruses as well (see references 5 and 14 for reviews). The literature specifically pertaining to VSV is reviewed here.

BINDING AND INTERNALIZATION OF VSV

The association of VSV with BHK cells is remarkably nonspecific. Figure 2 shows that, over a 300-fold range of VSV concentration, a nearly constant fraction of added virus bound to cells at 37°C; no evidence of saturation was observed (8). G protein, which is the only protein on the viral surface, is apparently not essential for binding to the cell surface, since (i) "spikeless" virions, resulting from proteolytic removal of G protein from the viral surface,

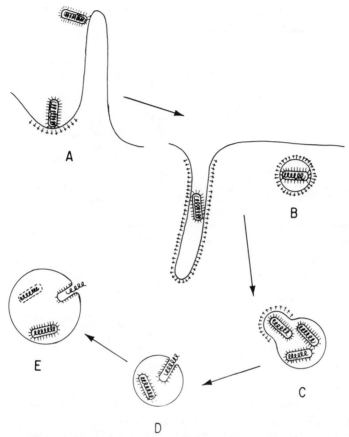

FIG. 1. Postulated sequence of events in VSV penetration and uncoating. (Adapted from reference 5.)

still bound to cells with about half the efficiency of intact virions (Fig. 2A) and (ii) thermal denaturation of the mutant G protein of the temperature-sensitive VSV mutant ts45 did not significantly decrease binding to cells (Fig. 2B; 8).

The extent of internalization of surface-bound virions is generally measured by their resistance to removal from the cell by degradative enzymes; internalized VSV cannot be removed from the cells by trypsin (8) or proteinase K (6) treatment. A nearly constant proportion of each bound virus was internalized at all virus concentrations studied, indicating that the internalization apparatus was not saturated (Fig. 2C and D). Internalization varied somewhat for the different viral preparations studied, ranging from 45 to 65%; removal or denaturation of G protein of VSV had minor effects on the efficiency of internalization of bound virions (Fig. 2C and D; 8).

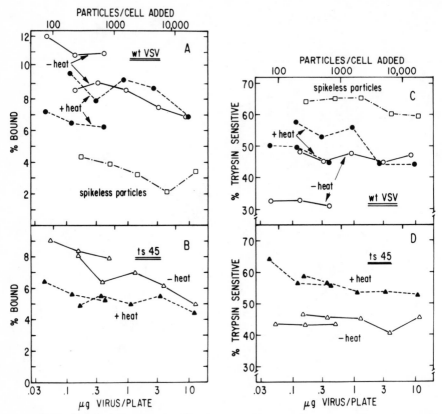

FIG. 2. Association of radiolabeled wild-type VSV (A) and temperature-sensitive VSV G protein mutant ts45 (B) with BHK cells after 45 min at 37°C as a function of the amount of virus added. (C and D) Amount of each virus remaining on the cell surface (trypsin sensitive) after 45 min at 37°C. (○) Wild-type VSV; (●) wild-type VSV heated at 45°C for 1 h before being added to cells; (□) spikeless particles (lacking G protein) prepared from wild-type VSV; (△) VSV ts45; (▲) VSV ts45 heated at 45°C for 1 h to denature its thermolabile mutant G protein before being added to cells. The results from two experiments are shown; points from each experiment are connected by lines. (From reference 8.)

A further indication that specific binding sites for VSV are not needed for uncoating or initiation of infection came from a study of the effect of the polycation DEAE-dextran. DEAE-dextran increased the binding of VSV to BHK cells about fourfold. The proportion of bound virions that was internalized remained the same; hence, about four times as many virions were internalized in the presence of DEAE-dextran as in its absence (1). Most significantly, the extent of primary viral transcription (RNA synthesis proceeding directly from

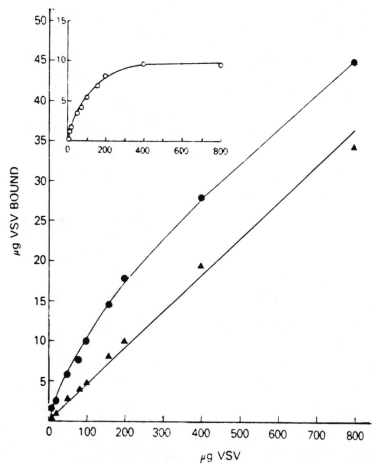

FIG. 3. Binding of radiolabeled VSV to Vero cells in the presence (▲) and absence (●) of excess unlabeled VSV after 20 h at 4°C. The inset shows the saturable binding (total binding minus nonsaturable binding). (From reference 12.)

input virions, in the absence of viral protein synthesis) increased by the same amount, about fourfold (1). The addition of DEAE-dextran evidently created new binding sites on the cell surface or augmented existing ones, and virions bound to these sites were fully capable of becoming uncoated and initiating a viral infection. Rabies virus, a virus related to VSV, showed similar enhancement by DEAE-dextran of binding to BHK cells and subsequent infection (16).

Although the precise chemical interactions of DEAE-dextran that result in enhanced VSV binding are not known, it seems likely that it functions to neutralize negative charge on the cell surface, thus facilitating binding of the

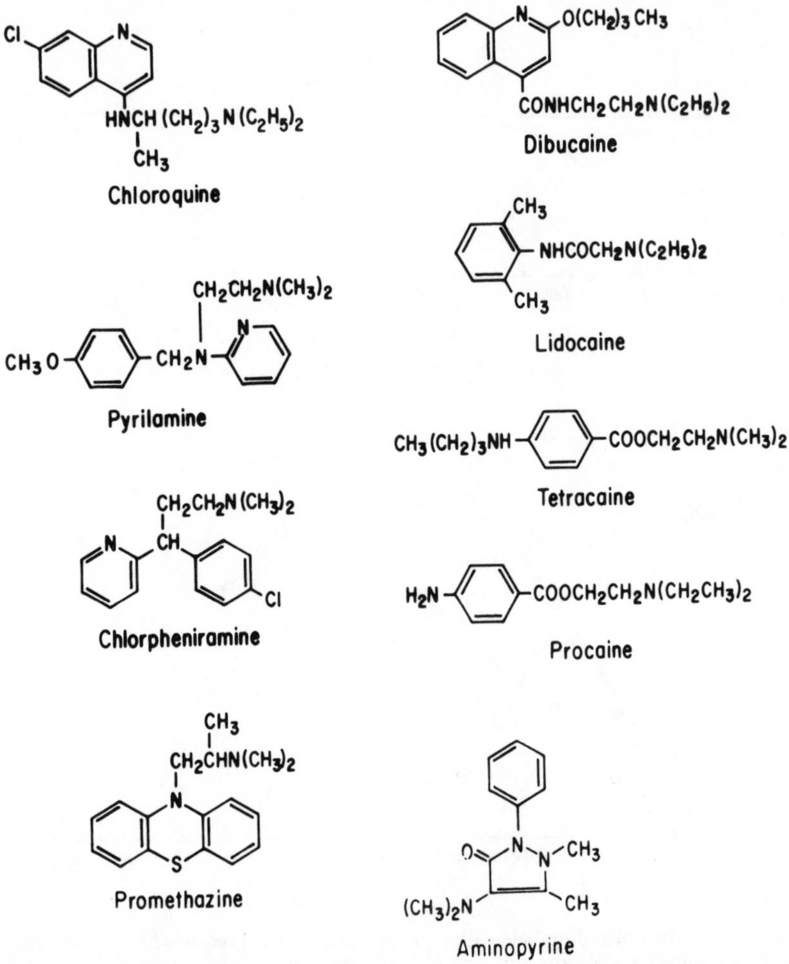

FIG. 4. Chemical structures of lysosomotropic agents of diverse pharmacologic activity that were tested (9) for ability to inhibit VSV infection of BHK cells.

negatively charged virion. Support for this idea comes from the observation that treatment of MDCK cells with neuraminidase results in a 2.5-fold enhancement of infection by VSV (4). The removal of sialic acid from the cell surface may resemble the addition of DEAE-dextran: both reduce cell surface negative charge.

In another study, Schlegel et al. found evidence for saturable binding sites for VSV on the surface of Vero cells at 4°C (12). (The results shown in Fig. 2 are from experiments performed at 37°C because over 80% of virions bound to BHK cells

TABLE 1. Inhibitors of VSV transcription[a]

Group	Drug	Concn for 50% inhibition (mM)
Antimalarial	Chloroquine	0.02
Local anesthetic	Dibucaine	0.025
	Tetracaine	0.20
	Lidocaine	0.22
	Procaine	3.3
Antihistaminic	Pyrilamine maleate	0.05
	Chlorpheniramine	0.4
	Promethazine hydrochloride	0.5
Antipyretic	Aminopyrine	2.3
Miscellaneous amines	Dansylcadaverine	0.4
	Ethylenediamine	1.5
	1-Propylamine	4
	Imidazole	4
	Methylamine	5.5

[a] From reference 9.

at 4°C became unbound within 10 min of incubation at 37°C; see Fig. 2a in reference 5.) The saturable sites are of relatively low affinity and high capacity (ca. 4,000 virions per cell) and are detected against a high background of nonsaturable binding (Fig. 3). Most importantly, the functional specificity of the saturable sites was not demonstrated; internalization appeared to occur equally well from saturable and nonsaturable sites (12). Similar saturable sites for rabies virus have been found on BHK cells (16). In agreement with the implications drawn from the effect of DEAE-dextran (1), the data in Fig. 3 therefore suggest that VSV may gain functional entry into the cell and be uncoated as a result of binding to at least two distinct classes of cell surface sites. Schlegel et al. have suggested that the phospholipid phosphatidylserine composes at least part of the saturable VSV binding site on the Vero cell surface (11).

UNCOATING BY ACID-MEDIATED FUSION

Internalized virions pass rapidly into endosomal vesicles (3, 6). Endosomes have been found to have acidic interiors, around pH 5 (13). If the model of VSV uncoating illustrated in Fig. 1 is correct, two expectations must be realized: (i) the virus envelope must fuse with other membranes at the endosomal pH (around 5), and (ii) raising endosomal pH must prevent uncoating and thereby inhibit virus infection. Both predictions have been experimentally tested.

Virus-induced hemolysis has proved to be a versatile and simple experimental approach to measure the fusion ability of viral envelopes. VSV is hemolytic to erythrocytes from several animal species, but only at pHs below 6 (1, 7). Maximal hemolytic activity is observed around pH 5.0. Interestingly, addition of DEAE-dextran or trypsinization of the erythrocyte surface, both of which

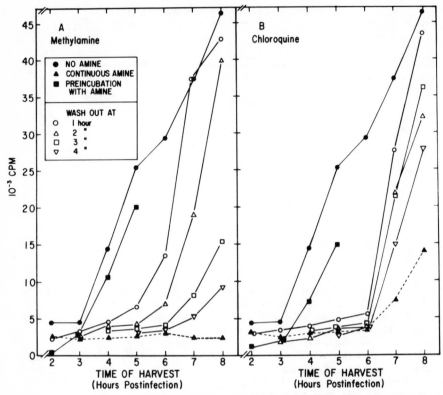

FIG. 5. Reversibility of inhibition of VSV RNA synthesis in BHK cells after removal of lysosomotropic amines. (A) Methylamine, 25 mM (from reference 9); (B) chloroquine, 0.1 mM.

decrease negative surface charge, often result in considerable enhancement of hemolytic activity (1, 7). VSV has also been shown to induce cell-cell fusion in monolayers by brief exposure of the cells, with bound virions attached, to pHs of 6 or lower (15). These observations establish the necessary fusogenic property of the VSV envelope, presumably mediated by the G protein, at the acidic pH found in endosomes.

A wide variety of organic amines diffuse across lipid membranes in their uncharged (basic) form and become trapped in acidic compartments after acquiring a proton (10). In this way, the acidic endosomes and lysosomes concentrate amines added to the cell medium, often several hundred-fold; the pH of the organelle rises to a new equilibrium value as a result, usually within 2 min (10). A characteristic acidic pH may be reestablished in the organelles by removing the amine from the cells, although recovery may take 1 h or more (10).

Most organic amines would behave as lysosomotropic agents, thus raising endosomal and lysosomal pH (2). A variety of organic amines, differing in both chemical structure and pharmacologic action, were therefore tested for their ability to inhibit VSV infection (9; Fig. 4). All of these amines were completely inhibitory at subtoxic concentrations (Table 1), consistent with the prediction that all effective lysosomotropic agents must act as inhibitors of viral uncoating (9). Other investigators found that ammonium chloride, another lysosomotropic agent, inhibited VSV infection (6). All these agents were effective only if they were present simultaneously with the input virions and for a relatively short time thereafter; addition even 30 min later greatly decreased their effectiveness as inhibitors. It seems reasonable to conclude that these diverse compounds all act in a similar way to disrupt an essential step early in viral infection.

The reversibility of the effect of lysosomotropic amines on VSV infection is shown in Fig. 5. If either methylamine or chloroquine was continuously present, no viral RNA synthesis occurred; however, pretreatment of the cells prior to addition of VSV had little or no effect. If the amine was present for only 1 or 2 h after inoculum removal, the reversibility of inhibition could be demonstrated. A rapid burst of RNA synthesis, eventually reaching control levels, commenced 1 or 2 h after removal of the amine. The longer lag time for chloroquine than for methylamine is consistent with a longer time required by cells for recovery from chloroquine (10). These results are consistent with the virions continuing to accumulate in the endosomes during the treatment period, followed by rapid uncoating once the endosomal pH was restored (9).

The pathway depicted in Fig. 1 has thus received strong experimental support as the predominant route of VSV uncoating in cultured cells.

Figure 5B depicts results of a previously unpublished experiment done in collaboration with Douglas K. Miller.

This work was supported in part by Public Health Service grant AI-13002 from the National Institute of Allergy and Infectious Diseases.

LITERATURE CITED

1. **Bailey, C. A., D. K. Miller, and J. Lenard.** 1984. Effects of DEAE-dextran on infection and hemolysis by VSV. Evidence that nonspecific electrostatic interactions mediate effective binding of VSV to cells. Virology **133**:111–118.
2. **DeDuve, C., T. deBarsy, B. Poole, A. Trouet, P. Tulkens, and F. van Hoof.** 1974. Lysosomotropic agents. Biochem. Pharmacol. **23**:2495–2531.
3. **Dickson, R. B., M. C. Willingham, and I. Pastan.** 1981. a₂-Macroglobulin adsorbed to colloidal gold: a new probe in the study of receptor-mediated endocytosis. J. Cell Biol. **89**:29–34.
4. **Griffin, J. A., S. Basak, and R. W. Compans.** 1983. Effects of hexose starvation and role of sialic acid in influenza virus release. Virology **125**:324–334.
5. **Lenard, J., and D. K. Miller.** 1983. Entry of enveloped viruses into cells, p. 119–138. *In* P. Cuatrecasas and T. F. Roth (ed.), Receptors and recognition, series B, vol. 15, Receptor-mediated endocytosis. Chapman & Hall, Ltd., London.
6. **Matlin, K. S., H. Reggio, A. Helenius, and K. Simons.** 1982. Pathway of vesicular stomatitis virus entry leading to infection. J. Mol. Biol. **156**:609–631.
7. **Mifune, K., M. Ohuchi, and K. Mannen.** 1982. Hemolysis and cell fusion by rhabdoviruses. FEBS Lett. **137**:293–297.
8. **Miller, D. K., and J. Lenard.** 1980. Inhibition of vesicular stomatitis virus infection by spike glycoprotein. Evidence for an intracellular, G protein-requiring step. J. Cell Biol. **84**:430–437.
9. **Miller, D. K., and J. Lenard.** 1981. Antihistaminics, local anesthetics and other amines as antiviral agents. Proc. Natl. Acad. Sci. USA **78**:3605–3609.
10. **Ohkuma, S., and B. Poole.** 1978. Fluorescence probe measurement of the intralysosomal pH in living cells and the perturbation of pH by various agents. Proc. Natl. Acad. Sci. USA **75**:3327–3331.

11. **Schlegel, R., T. S. Tralka, M. C. Willingham, and I. Pastan.** 1983. Inhibition of VSV binding and infectivity by phosphatidylserine: is phosphatidylserine a VSV-binding site? Cell **32:**639–646.
12. **Schlegel, R., M. C. Willingham, and I. H. Pastan.** 1982. Saturable binding sites for vesicular stomatitis virus on the surface of Vero cells. J. Virol. **43:**871–875.
13. **Tycko, B., and F. R. Maxfield.** 1982. Rapid acidification of endocytic vesicles containing a_2-macroglobulin. Cell **28:**643–651.
14. **White, J., M. Kielian, and A. Helenius.** 1983. Membrane fusion proteins of enveloped animal viruses. Q. Rev. Biophys. **16:**151–195.
15. **White, J., K. Matlin, and A. Helenius.** 1981. Cell fusion by Semliki Forest, influenza and vesicular stomatitis viruses. J. Cell Biol. **89:**674–679.
16. **Wunner, W. H., K. J. Reagan, and H. Koprowski.** 1984. Characterization of saturable binding sites for rabies virus. J. Virol. **50:**691–697.

Pathway of Adenovirus Entry into Cells

PREM SETH, DAVID FITZGERALD, MARK WILLINGHAM,
AND IRA PASTAN

*Laboratory of Molecular Biology, Division of Cancer Biology and Diagnosis,
National Cancer Institute, Bethesda, Maryland 20892*

**Many viruses, growth factors, hormones, and nutrients enter cells
by receptor-mediated endocytosis. In this pathway the ligands enter
coated pits and soon find themselves in intracellular vesicles that
have been formed from the coated pits. These vesicles are termed
receptosomes or endosomes and have an acid pH. Adenovirus and
some other viruses escape from the endocytic vesicle in a reaction
facilitated by low pH. Adenovirus does this by disrupting the
vesicle, releasing the virus and other contents of the vesicle into the
cytoplasm. Adenovirus entry into the cytoplasm is inhibited by
treatment of the virus with antibodies to the penton base protein. At
pH 5.5 adenovirus will directly permeabilize the plasma membrane
prior to its entry into cells by endocytosis. This results in the release
from cells of choline, ^{51}Cr, deoxyglucose, and α-aminobutyric acid,
but not large molecules. The hydrophobic properties of the capsid
proteins have been studied by measuring their interaction with
Triton X-114. The penton base strongly binds to Triton X-114 at pH
5.5 and below, and weakly binds at pH 7 and above. These findings
implicate the penton base protein as having an important role in
vesicle disruption.**

Recently, interest in how adenovirus enters cells has been rekindled (2, 3, 17,
20). In early studies, when adenovirus entry was followed by using electron
microscopy, virions were often seen in membrane-limited vacuoles shortly after
they were internalized from the cell surface (2). Other investigators, however,
have presented evidence that the majority of adenovirus particles enter cells by
direct penetration through the plasma membrane (11). Our group has been
studying the pathway of adenovirus entry for just a few years (4). Our data
suggest that adenovirus enters into cells by receptor-mediated endocytosis; the
early steps involved in the uptake of adenovirus are quite similar to those of
many other ligands known to be internalized by receptor-mediated endocytosis
(6, 7, 13, 14). In the following sections, we describe experiments directed at
understanding how adenovirus enters into the cytoplasm and what the basic
mechanisms involved in this process might be.

ENTRY OF ADENOVIRUS INTO THE CYTOPLASM

It has been suggested that adenovirus binds to the cell surface through its
fiber protein, because purified fiber can inhibit the attachment of adenovirus to
cells (16). Using this competition assay, Philipson et al. estimated that a single

191

KB cell has about 10,000 binding sites for adenovirus (16). Since then, a few attempts have been made to identify the receptor for adenovirus (1, 8, 21). Although the exact nature of the receptor molecule is still under study, it appears that more than one plasma membrane protein might be involved in the adenovirus-receptor interaction. After binding to the cell surface, adenovirus bound to its receptor next appears in clathrin-coated pits (2, 4). Coated pits are specialized depressions in the plasma membrane that function to concentrate ligands bound to the cell surface and serve as an entry site for selective endocytosis. Adenovirus particles are next seen in uncoated endocytic vesicles. These endocytic vesicles are isolated intracellular structures which were earlier called vacuoles (2) and now are termed receptosomes (15) or endosomes (7). Adenovirus may enter into the cytoplasm by rupturing the membrane of the receptosome (2, 4). Some particles that do not escape from receptosomes can later be found in lysosomes.

ESCAPE OF ADENOVIRUS FROM RECEPTOSOMES: ROLE OF LOW pH

The selective rupture of receptosomes by adenovirus is suggested by certain experimental results. If adenovirus ruptures the receptosome during its entry into the cytoplasm, any other molecule(s) present in the receptosome along with the adenovirus might also be released into the cytoplasm. When we allowed a cell to cointernalize adenovirus and epidermal growth factor (EGF) linked to colloidal gold, we found gold particles released into the cytoplasm only when adenovirus was present (4). In another experiment EGF was linked to *Pseudomonas* exotoxin (EGF-PE), and both virus and toxin appeared to enter through a common receptosome. Normally, in the absence of virus, very little EGF-PE escapes into the cytoplasm; instead it is delivered to and degraded in the lysosomes. Adenovirus causes an increase in the release of EGF-PE into the cytoplasm and results in the enhancement of the toxicity of EGF-PE. EGF-PE has the same mechanism of action as native PE. It ADP-ribosylates elongation factor 2 and thus inhibits protein synthesis. We have used the enhancing activity of adenovirus as a measure of the ability of the virus to rupture receptosomes and escape into the cytoplasm.

One characteristic feature of receptosomes is that they have an acidic interior with a pH of about 5 to 5.5 (22). In the presence of various weak bases including chloroquine, which raise the pH of the receptosome to near neutrality (9), the rupture of receptosomes by adenovirus was significantly inhibited (17). It was, therefore, concluded that the low pH of receptosomes is important for adenovirus to escape from receptosomes into the cytoplasm. One possible effect of low pH on adenovirus might be to increase its ability to interact with the membrane of the receptosome by exposing hydrophobic residues on the proteins of the virus surface. With this in mind, we tested the hydrophobicity of adenovirus and its external proteins at different pH values by measuring their ability to interact with the nonionic detergent Triton X-114 (P. Seth, M. C. Willingham, and I. Pastan, submitted for publication). It was found that the ability of adenovirus to associate with Triton X-114 was much greater at pH 5.0 than at neutral pH. Among the three external proteins (hexon, penton base, and fiber), penton base associated most with Triton X-114 at pH 5.0. On the basis of

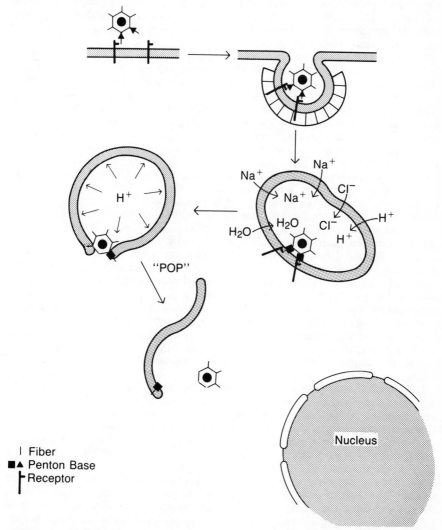

FIG. 1. Model of adenovirus entry and vesicle disruption. Adenovirus binds to a cell surface receptor and moves into a coated pit. Soon thereafter it appears in an uncoated endocytic vesicle termed a receptosome (endosome). The low pH of the receptosome causes the external proteins of adenovirus, especially the penton base protein, to undergo a conformational change (▲ → ■). As a result of this, adenovirus acquires amphiphilic characteristics which enable it to interact with the membrane of the receptosome. The receptosome, which is distended as a result of osmotic pressure generated by ion pumps in its membrane, ruptures at the point where adenovirus penetrates the membrane.

these results it is suggested that mildly acidic pH induces amphiphilic properties in the adenovirus capsid proteins, perhaps by exposing hidden hydrophobic
sites. As a result of this, adenovirus can interact with the lipid components of
the receptosome membrane.

We are also interested in understanding how the low-pH-induced hydrophobic interactions of adenovirus with the receptosome membranes cause them to
rupture. Since this reaction occurs intracellularly, it has been difficult to study
the mechanism directly. Therefore, we developed a more convenient way to
study adenovirus effects on cell membranes. Adenovirus was first bound to the
cell surface, and subsequently the pH of the medium was lowered to 6.0. During
the low-pH incubation, there was an adenovirus-dependent increase in cell
membrane permeability. This was measured by following the release of small
molecules such as ^{51}Cr, [^{14}C]choline, α-aminobutyric acid, and 2-deoxyglucose
(19; P. Seth, I. Pastan, and M. C. Willingham, J. Biol. Chem., in press). The
biochemical nature of this adenovirus effect on the cell surface appears to share
several common features with its effect on receptosomes. For example, (i) the
adenovirus effect on cell membrane permeability is favored when the virus is
incubated with cells at mildly acidic pH (~6); receptosome rupture is favored
when adenovirus is present in the normal acidic environment of the receptosome. (ii) Both of the adenovirus effects show a strong temperature dependence,
as one expects in a reaction involving membrane perturbation. (iii) Both of the
effects of adenovirus are completely prevented when adenovirus is inactivated
at 45°C for 10 min (Seth et al., in press; unpublished data), suggesting a
requirement for native conformation of adenovirus protein(s) in these interactions. (iv) Both of the effects are specifically blocked by a polyclonal antiserum
against penton base protein, suggesting the involvement of this protein (18). It
is conceivable, therefore, that cell surface effects of adenovirus on the plasma
membrane might be due to an activity which adenovirus normally uses to
rupture acidic endocytic vesicles.

It is possible that adenovirus cannot produce more than local lesions in the
cell membrane because it is closely associated with an extensive submembran-
ous cytoskeletal network which is thought to stabilize its integrity (24).
Endocytic vesicles are not associated with this actin meshwork and are easily
disrupted by osmotic rupture, whether present intracellularly (3, 12) or
extracellularly after purification from cell homogenates (3). Whether rupture or
lysis of receptosomes by adenovirus is preceded by a change in the permeability
across the receptosome membrane is not known. In this regard, it is important
to note that receptosome formation and acidification probably involve proton
pumps and other ion pumps (5, 10, 23). Adenovirus could affect the activity of
these pumps, causing a perturbation of the osmotic pressure within the vesicles.
As a result of this, receptosomes might rupture and release adenovirus particles
into the cytoplasm. A schematic representation of some of the events in
adenovirus entry is shown in Fig. 1.

LITERATURE CITED

1. **Butters, T. D., and R. C. Hughes.** 1984. Solubilization and fractionation of glycoproteins and
   glycolipids of KB cell membranes. Biochem. J. **140:**469–478.
2. **Dales, S.** 1973. Early events in cell-animal virus interactions. Bacteriol. Rev. **37:**103–135.

3. **Dickson, R. B., L. Beguinot, J. A. Hanover, N. D. Richert, M. C. Willingham, and I. Pastan.** 1983. Isolation and characterization of a highly enriched preparation of receptosomes (endosomes) from a human cell line. Proc. Natl. Acad. Sci. USA **80:**5335–5339.
4. **FitzGerald, D. J. P., R. Padmanabhan, I. Pastan, and M. C. Willingham.** 1983. Adenovirus-induced release of epidermal growth factor and Pseudomonas toxin into the cytosol of KB cells during receptor-mediated endocytosis. Cell **32:**607–617.
5. **Galloway, C. J., G. E. Dean, M. Marsh, G. Rudnick, and I. Mellman.** 1983. Acidification of macrophage and fibroblast endocytic vesicles *in vitro*. Proc. Natl. Acad. Sci. USA **80:**3334–3338.
6. **Goldstein, J. L., R. G. W. Anderson, and M. S. Brown.** 1979. Coated pits, coated vesicles, and receptor-mediated endocytosis. Nature (London) **279:-**679–685.
7. **Helenius, A., I. Mellman, D. Wal, and A. Hubbard.** 1983. Endosomes. Trends Biochem. Sci. **8:**245–250.
8. **Hennache, B., and P. Boulanger.** 1977. Biochemical study of KB-cell-receptor for adenovirus. Biochem. J.**166:**237–247.
9. **Maxfield, F. R.** 1982. Weak bases and ionophores rapidly and reversibly raise the pH of endocytic vesicles in cultured mouse fibroblasts. J. Cell Biol. **95:**676–681.
10. **Merion, M., P. Schlesinger, R. M. Brooks, J. M. Moehring, T. J. Moehring, and W. S. Sly.** 1983. Defective acidification of endosomes in Chinese hamster ovary cell mutants "cross-resistant" to toxins and viruses. Proc. Natl. Acad. Sci. USA **80:**5315–5319.
11. **Morgan, C., H. S. Rosenkranz, and B. Mednis.** 1969. Structure and development of viruses as observed in the electron microscope. X. Entry and uncoating of adenovirus. J. Virol. **4:**777–796.
12. **Okada, C. Y., and M. Rechsteiner.** 1982. Introduction of macromolecules into cultured mammalian cells by osmotic lysis of pinocytic vesicles. Cell **29:**33–41.
13. **Pastan, I., and M. C. Willingham.** 1981. Journey to the center of the cell: role of the receptosome. Science **214:**504–509.
14. **Pastan, I. H., and M. C. Willingham.** 1981. Receptor-mediated endocytosis of hormones in cultured cells. Annu. Rev. Pysiol. **43:**239–250.
15. **Pastan, I., and M. C. Willingham.** 1983. Receptor-mediated endocytosis: coated pits, receptosomes, and the Golgi. Trends Biochem. Sci. **8:**250–254.
16. **Philipson, L., K. Lonberg-Holm, and V. Peterson.** 1968. Virus-receptor interaction in an adenovirus system. J. Virol. **2:**1064–1075.
17. **Seth, P., D. J. P. FitzGerald, M. C. Willingham, and I. Pastan.** 1984. Role of a low-pH environment in adenovirus enhancement of the toxicity of a *Pseudomonas* exotoxin-epidermal growth factor conjugate. J. Virol. **51:**650–655.
18. **Seth, P., D. FitzGerald, H. Ginsberg, M. Willingham, and I. Pastan.** 1984. Evidence that the penton base on adenovirus is involved in potentiation of toxicity of *Pseudomonas* exotoxin conjugated to epidermal growth factor. Mol. Cell. Biol. **4:**1528–1533.
19. **Seth, P., M. C. Willingham, and I. Pastan.** 1984. Adenovirus dependent release of ^{51}Cr from KB cells at an acidic pH. J. Biol. Chem. **259:**14350–14353.
20. **Svensson, U., and R. Persson.** 1984. Entry of adenovirus 2 into HeLa cells. J. Virol. **51:**687–697.
21. **Svensson, U., R. Persson, and E. Everitt.** 1981. Virus-receptor interaction in the adenovirus system. I. Identification of virion attachment proteins of the HeLa cell plasma membrane. J. Virol. **38:**70–81.
22. **Tycko, B., and F. R. Maxfield.** 1982. Rapid acidification of endocytic vesicles containing α_2-macroglobulin. Cell **28:**643–651.
23. **Willingham, M. C., and I. Pastan.** 1983. Formation of receptosomes from plasma membrane coated pits during endocytosis: analysis by serial sections with improved membrane labeling and preservation techniques. Proc. Natl. Acad. Sci. USA **80:**5617–5621.
24. **Willingham, M. C., S. S. Yamada, P. J. A. Davies, A. V. Rutherford, M. G. Gallo, and I. Pastan.** 1981. The intracellular localization of actin in cultured fibroblasts by electron microscopic immunocytochemistry. J. Histochem. Cytochem. **29:**17–37.

Requirements for Initial Adenovirus Uncoating and Internalization

EINAR EVERITT, ULLA SVENSSON, CLAES WOHLFART, AND ROBERT PERSSON

Department of Microbiology, The University of Lund, S-223 62 Lund, Sweden

The early extracellular interaction between adenovirus type 2 and permissive HeLa cells was studied biochemically and with an electron microscope. Attachment in the presence of dithiothreitol or dansylcadaverine, or attachment at different temperatures, suggested that diffusion in the plane of the plasma membrane provided the basis for an observed positive cooperative binding. Cooperativity depended upon the multivalency of the virion's attachment proteins, i.e., the fibers. Inhibitors of cooperativity also affected subsequent uncoating, suggesting that precise microenvironmental conditions are required for this process. If cells were treated with the two reagents above, or if virions were fixed with glutaraldehyde, virions accumulated within vesicles in the infected cell. This suggested that virion destabilization, previously shown to occur at the cell surface, is a prerequisite for escape of virions from the intracellular vesicles on their route to the nucleus. Possible involvement of enzymatic events and effects of membrane fluidizers in attachment and uncoating are also discussed.

In most virus-eucaryotic cell systems the PFU-to-virion ratio varies considerably, and for different serotypes of the *Adenoviridae* family a variation in specific infectivity between 0.04 and 10% has been observed (10). This must always be considered when results based on electron microscopic and biochemical analyses are interpreted regarding the cellular localization and the physical status of infecting virus particles. The initial steps of adenovirus interaction with permissive cells have been studied by such techniques, and at present it appears as if adenoviruses of different serotypes, and within the same serotype, may force the plasma membrane barrier by two alternative mechanisms. One mechanism is by direct penetration (3, 18), and the other is by a "phagocytotic" (4, 18) or an endocytotic process (4, 9, 31). These variations could be reflected in observed differences in cellular distribution of infecting virions in different systems (5). In the following discussion the various serotypes will not be treated separately.

Whether adenovirus virions display any morphological changes soon after internalization is a matter of controversy. Thus, some investigators claim that virions appear more rounded and evidently have lost some of their icosahedral appearance early after penetration (3, 18). Other studies describe unaltered morphologies of the penetrated inoculum (15, 19, 29), although virions definitely become distorted when localized in the vicinity of the nuclear membrane after prolonged incubation (15). The problems of detecting subtle morphological changes in infecting virions by electron microscopy have been thoroughly discussed by Dales (6). A number of biochemical investigations support the solid statement that adenoviruses in association with the plasma membrane at the cellular surface are destabilized so that components of the viral vertex regions are released (2, 4, 14, 15, 24, 29). A possibly analogous labilization of the vertices of adenovirus type 2 (Ad2) is easily achieved upon dialysis of purified virus against hypotonic buffers of low pH (8, 25). As a consequence of such destabilization, the virus infectivity disappears, and its disappearance is paralleled by sensitization of the parental genome to exogenously added DNase (25). This situation probably mimics the initial step of uncoating, and in both systems about 80% of the genome becomes DNase sensitive (13, 23, 25, 31).

In our laboratory we have reexamined some of the very early events of the interaction between Ad2 and the permissive HeLa cell. The present knowledge of other ligand-receptor systems and systems of receptor-mediated endocytosis suggested testing the effects of drugs and reagents and of a wide temperature range on the various substeps leading to uncoating and internalization of Ad2.

ATTACHMENT AND COOPERATIVE BINDING

An interesting feature emerging from a series of simple experiments was that the temperature dependence of the virus attachment rate increased abruptly at ca. 20°C (21). A possible hint of a mechanism behind this temperature-dependent shift was obtained when attachment of Ad2 to cells was studied over a broad temperature range and at multiplicities of infection between 50 and 20,000 particles per cell (22). (In the course of this work we experimented with different transformations of our data. Although attachment of adenovirus to host cells is not a reversible process, and no true equilibrium is ever established, we found it useful to calculate a "binding constant" from our attachment data by the method of Scatchard [27]. This procedure was established only for reversible binding of univalent ligands to receptors. We therefore consider our binding constants as empirical rather than as absolute. The attachment data have also been transformed by the method of Hill [12], and empirical Hill coefficients [N_E, Table 1] were calculated from fractional binding observed at different virus multiplicities, as described by Persson et al. [22]. For this transformation, the fractional occupancy of cellular receptor sites was calculated using for full occupancy the number of virions attached with saturating levels of input virus and with other conditions such as temperature or concentration of inhibitor constant. Again, since adenovirus attachment to host cells is an irreversible system, our coefficients have no strict physical meaning, but seem intuitively to be useful for recognition of cooperative binding.)

Binding of virions at temperatures between 5 and 20°C occurred with increasing positive cooperativity (22). The positive cooperativity was totally inhibited by prefixation of cells with 0.015% glutaraldehyde and at temperatures around 5°C, suggesting that free migration was required for components of the plasma membrane. At temperatures above 20°C the positive cooperativity reached a maximum (Fig. 1), which was maintained up to 37°C. Taken together, the most immediate interpretation of these observations suggests that lateral movement of virus-receptor complexes contributes to positive cooperativity

TABLE 1. Manipulations of the steps of cooperative binding, uncoating, and
penetration in the Ad2 system

Treatment of cells	Relative cooperativity (%)[a]	Relative uncoating (%)[b]	Relative penetration (%)[b]	Relative cellular localization (%)[c]		
				Surface	Vesicles	Free in cytoplasm
Control, 37°C	100	100	100	12	8	80
3°C	0	0	0	95	5	0
50 mM sodium azide	100	75	0	84	16	0
10 mM dithiothreitol	17	50	80	36	32	32
5 mM EDTA	100	—[d]	—	72	13	16
20 mM EDTA	100	30	25	—	—	—
20 mM EGTA	95	44	15	—	—	—
1 mM dansyl-cadaverine	49	3	0	53	44	3

[a] Data obtained from Persson et al. (22). The relative cooperativity was calculated by the formula: relative cooperativity = $[N_{E(\text{sample})} - N_{E(3°C)}]/[N_{E(37°C)} - N_{E(3°C)}] \times 100$, where N_E is the empirical Hill coefficient derived from the binding data (22). See the text.

[b] Data obtained from Svensson and Persson (31).

[c] Data obtained from Svensson (30).

[d] —, Not done.

and occurs at the same temperature as hypothetical local phase transitions or phase separations. Consequently, the initial attachment rates increase at this "melting" temperature. The positive cooperativity at 37°C was not affected by sodium azide (22; Table 1), an inhibitor of oxidative phosphorylation, suggesting that the positive cooperativity may be a result of a patching or microaggregation process and is separate from capping (21) and redistributional phenomena (20). One established cause of positive cooperativity is multivalency of the ligand. In the adenovirus system the 12 virus attachment proteins, i.e., the fiber structures, account for this. Thus, multiplicity-dependent attachment studies with purified Ad2 fiber antigen revealed poor or no positive cooperativity irrespective of the temperature (22).

UNCOATING AND COOPERATIVITY

We will refer to the early process of virus destabilization, rendering the DNA core sensitive to DNase treatment, as initial uncoating. A compilation of our previous and more recent data in Table 1 suggested a possible relationship between positive cooperativity and the initial step of uncoating, both processes believed to occur on the cell surface. Thus, both events were totally blocked at low temperatures, and the metabolic inhibitor sodium azide displayed no effect and a 25% reduction on the cooperativity and initial uncoating, respectively. In the presence of azide the penetration step was inhibited as measured by a technique using radiolabeled anti-Ad2 antibodies (31). Recent studies with an electron microscope corroborate these data (20, 30).

In a number of biological systems the affinity and functional activity of membrane receptors are regulated by the local microviscosity. For example,

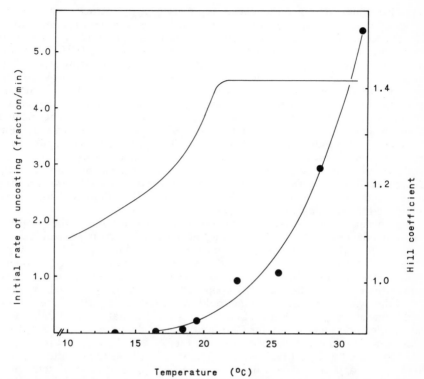

FIG. 1. Temperature dependence of the positive cooperativity of Ad2 attachment ᴄᴏ HeLa cells and of the initial rate of uncoating. The initial uncoating rates were measured as previously described (31), but with a change in the protocol for the cell disruptions, which were performed in the presence of 0.5% deoxycholate. Cells were washed once in phosphate-buffered saline and chilled on ice for 15 min at a density of 5×10^7 cells per ml. [³H]thymidine-labeled Ad2 (71,900 cpm/10^{10} virions) was added at a multiplicity of infection of 1,500 and attached at 3 to 6°C for 15 min to give an attached multiplicity of about 150 (21). Unadsorbed virus was removed after dilution six times in ice-cold phosphate-buffered saline and subsequent sedimentation of cells. The cells were resuspended in phosphate-buffered saline to give 2.5×10^7 cells per ml and incubated at the appropriate temperatures, and samples of 200 µl were removed at 5-min intervals for 20 min. The curve with the symbol ● depicts the extent of uncoating as the percentage of DNA rendered sensitive to DNase treatment per unit of time. The temperature dependence of the positive cooperativity (solid line; see the text) is taken from Persson et al. (22), where an empirical Hill coefficient above unity indicates a positive cooperativity.

aliphatic alcohols were used to study chemotactic factor receptors on human polymorphonuclear leukocytes (32). Among these so-called membrane fluidizers, the local anesthetic tetracaine and other small amphipathic molecules have been investigated. Apart from fluidizing properties (26), these molecules may also exhibit detergentlike effects and consequently extract proteins from the membrane (16). Experiments were designed to see whether the temperature dependence of attachment cooperativity and uncoating was directly coupled to the fluidity of the plasma membrane. Since primary amines (NH_4Cl, chloroquine, or methylamine) did not seriously affect the initial uncoating of Ad2 (31), tetracaine, a lipophilic amine membrane perturber (17), was chosen. Virus-cell binding studies were performed at different multiplicities of infection, at four temperatures, and in the presence of one selected concentration of tetracaine. By comparison with each control series, it was found that 0.5 mM tetracaine caused a significant reduction in the number of receptors and also a reduction in the empirical binding constant (not shown). These changes may, in part, have resulted from the presence of this detergentlike substance during virus attachment. Besides these negative effects, a small (about 10%) but reproducible increase of the positive cooperativity was seen at the intermediate temperatures of 10.5 and 14°C, suggesting that the presence of the "fluidizer" may have enhanced lateral diffusion of virus-receptor complexes.

The uncoating step was subsequently studied. Figure 1 shows the temperature dependence of the initial rates of adenovirus uncoating in control incubations compared with the positive cooperativity of binding obtained from Persson et al. (22). From these curves it appears that temperature-dependent enhancement of uncoating rates is most pronounced at temperatures above approximately 20°C, where positive cooperativity is already maximal. We then tested the hypothesis that tetracaine could counteract the rigidity of the plasma membrane at low temperatures. Control tests showed that virion stability was not affected upon incubation for 60 min at 37°C in the presence of 0.5 and 1.0 mM tetracaine. The next set of experiments tested whether 0.5 mM tetracaine could increase the initial rate of uncoating, specifically at temperatures below 20°C. Quite opposite from what was expected, the presence of tetracaine significantly enhanced the relative initial rate of uncoating at 20°C and above (Fig. 2). The rate enhancement gradually diminished with temperatures above 20°C and approached nil at temperatures above 30°C where maximum uncoating rates were achieved (31). Since the effects of tetracaine on uncoating were obvious only at temperatures above 20°C and where positive cooperativity was already maximal, the anticipated fluidizing effects may have provided increased enzymatic activities due to local microenvironmental fluidity changes of the plasma membrane (26), rather than enhanced lateral diffusion of virus-receptor complexes.

UNCOATING AND INTERNALIZATION

The chelators EDTA and ethylene glycol-bis(β-aminoethyl ether)-N,N,N',N',-tetraacetic acid (EGTA) both affected uncoating and internalization (Table 1), suggesting the involvement of divalent cations in these events. Among agents previously known to impair migration of ligand-receptor complexes (11, 28), the disulfide reducing agent dithiothreitol and the inhibitor of endocytosis dansylcadaverine both inhibited uncoating and internalization. The mechanisms behind these effects are open to speculation, but interference with transglutaminases is a possibility (7).

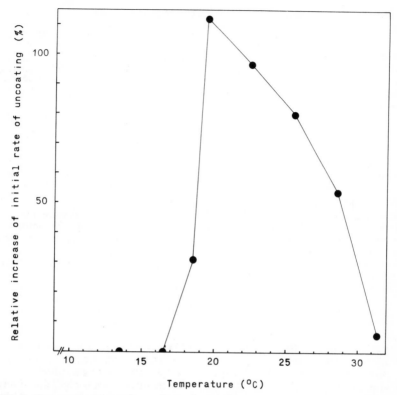

FIG. 2. Temperature dependence of initial rates of uncoating in the presence of tetracaine. Assays for the initial rates of uncoating were performed as described in the legend of Fig. 1. After virus attachment, dilution, and sedimentation, one half of the cells were suspended in phosphate-buffered saline as the control series, and the other half were suspended in phosphate-buffered saline containing 0.5 mM tetracaine. The increase of initial rates of uncoating for the tetracaine series relative to the control series is plotted versus the temperature of incubation. Absolute uncoating data for the control series are shown in Fig. 1.

The data of Table 1 suggested that intravesicular accumulation of the infecting inoculum may be a consequence of internalization of virions which have avoided uncoating. Accordingly, virions artificially stabilized with 0.015% glutaraldehyde were totally unable to escape from the vesicular environment (Table 2). Such effects have previously been reported for formaldehyde-treated virus (18) and for virus treated by heating at both 45 and 60°C (3).

Manipulation of virion topography by attachment of different classes of neutralizing antibodies should also interfere with the uncoating and internalization processes. Virions treated with monospecific antihexon and antifiber antibodies, at serum-to-virus ratios yielding complete neutralization, showed a significant intravesicular and surface accumulation, respectively (Table 2). Inter-

TABLE 2. Early interaction of modified Ad2[a]

Treatment of virus	Relative uncoating (% of control)[b]	Relative neutralization (% of control)[c]	Relative cellular localization[d] (% of total cell-associated particles)		
			Surface	Vesicles	Free in cytoplasm
Glutaraldehyde fixation	0	—	15	85	0
Ad2 + antihexon antibodies, 1:1.25	100	100	20	70	10
Ad2 + antifiber antibodies, 1:200	26	100	85	15	0
Preimmune serum, 1:1.25	100	0	16	16	68

[a] From Wohlfart et al. (34).
[b] After attachment to monolayer cell cultures of 2,000 virions or virion-antibody complexes per cell, the uncoating efficiencies were related to an appropriate control value after 60 min of combined attachment and uncoating incubation at 37°C. The DNase sensitivity assays were performed as previously described (31).
[c] Virus neutralization was assayed as reduction in the amount of progeny virus synthesized in suspension cultures (33).
[d] The relative cellular localization of virions was scored in an electron microscope after 60 min of attachment at 37°C. Virions or virion-antibody complexes were added at multiplicities of infection of 5,000, and each sample series contained 5×10^7 cells.

estingly, virus neutralized by antihexon antibodies displayed normal uncoating, and these virions could not escape from the vesicular compartment, as observed in an electron microscope, probably because they were coated by antibodies.

It appears that malfunctional virus is distributed to vesicles, and further characterization of these vesicles should be interesting, especially in the light of the fact that isolated lysosomal enzymes alone do not seem to be responsible for the process of adenovirus uncoating (1).

CONCLUDING REMARKS

At a temperature where maximum positive attachment cooperativity is achieved, efficient uncoating first appears to occur. The temperature dependence of the positive cooperativity of attachment may be linked to microaggregational events (clustering) needed before initiation of the uncoating step. The internalization step may be enzymatically governed, because it is inhibited by azide and by a transglutaminase inhibitor and it requires divalent cations. Under conditions which inhibit uncoating, virions tend to accumulate in intracellular vesicles. This is also the case when virus which has chemically cross-linked capsids or which has reacted with neutralizing antihexon antibodies is permitted to interact with cells.

The excellent technical assistance of Blanka Boberg is much appreciated.
This investigation was financially supported by the Swedish Natural Science Research Council.

LITERATURE CITED

1. **Boulanger, P. A., M. D. Breynaert, and G. Biserte.** 1970. Lysosomes and the problem of adenovirus uncoating. Exp. Mol. Pathol. **12:**235–242.
2. **Boulanger, P. A., and B. Hennache.** 1973. Adenovirus uncoating: an additional evidence for the involvement of cell surface in capsid labilization. FEBS Lett. **35:**15–18.
3. **Brown, D. T., and B. T. Burlingham.** 1973. Penetration of host cell membranes by adenovirus 2. J. Virol. **12:**386–396.
4. **Chardonnet, Y., and S. Dales.** 1970. Early events in the interaction of adenoviruses with HeLa cells. I. Penetration of type 5 and intracellular release of the DNA genome. Virology **40:**462–477.
5. **Chardonnet, Y., and S. Dales.** 1970. Early events in the interaction of adenoviruses with HeLa cells. II. Comparative observations on the penetration of types 1, 5, 7, and 12. Virology **40:**478–485.
6. **Dales, S.** 1973. Early events in cell-animal virus interactions. Bacteriol. Rev. **37:**103–135.
7. **Davies, P. J. A., D. R. Davies, A. Levitzki, F. R. Maxfield, P. Milhaud, M. C. Willingham, and I. H. Pastan.** 1980. Transglutaminase is essential in receptor-mediated endocytosis of α_2-macroglobulin and polypeptide hormones. Nature (London) **283:**162–167.
8. **Everitt, E., B. Sundquist, U. Pettersson, and L. Philipson.** 1973. Structural proteins of adenoviruses. X. Isolation and topography of low molecular weight antigens from the virion of adenovirus type 2. Virology **52:**130–147.
9. **FitzGerald, D. J. P., R. Padmanabhan, I. Pastan, and M. C. Willingham.** 1983. Adenovirus-induced release of epidermal growth factor and Pseudomonas toxin into the cytosol of KB cells during receptor-mediated endocytosis. Cell **32:**607–617.
10. **Green, M., M. Piña, and R. C. Kimes.** 1967. Biochemical studies on adenovirus multiplication. XII. Plaquing efficiencies of purified human adenoviruses. Virology **31:**562–565.
11. **Hazum, E., K.-J. Chang, and P. Cuatrecasas.** 1979. Role of disulphide and sulfhydryl groups in clustering of enkephalin receptors in neuroblastoma cells. Nature (London) **282:**626–628.
12. **Hill, A. V.** 1910. The possible effects of the aggregation of the molecules of haemoglobin on its dissociation curves. J. Physiol. (London) **40:**iv–vii.
13. **Lawrence, W. C., and H. S. Ginsberg.** 1967. Intracellular uncoating of type 5 adenovirus deoxyribonucleic acid. J. Virol. **1:**851–867.
14. **Lonberg-Holm, K., and L. Philipson.** 1969. Early events of virus-cell interaction in an adenovirus system. J. Virol. **4:**323–338.
15. **Lyon, M., Y. Chardonnet, and S. Dales.** 1978. Early events in the interaction of adenoviruses with HeLa cells. V. Polypeptides associated with the penetrating inoculum. Virology **87:**81–88.
16. **Maher, P., and S. J. Singer.** 1984. Structural changes in membranes produced by the binding of small amphipathic molecules. Biochemistry **23:**232–240.
17. **Miller, D. K., and J. Lenard.** 1981. Antihistaminics, local anesthetics, and other amines as antiviral agents. Proc. Natl. Acad. Sci. USA **78:**3605–3609.
18. **Morgan, C., H. S. Rosenkranz, and B. Mednis.** 1969. Structure and development of viruses as observed in the electron microscope. X. Entry and uncoating of adenovirus. J. Virol. **4:**777–796.
19. **Ogier, G., Y. Chardonnet, and W. Doerfler.** 1977. The fate of type 7 adenovirions in lysosomes of HeLa cells. Virology **77:**67–77.
20. **Patterson, S., and W. C. Russell.** 1983. Ultrastructural and immunofluorescence studies of early events in adenovirus-HeLa cell interaction. J. Gen. Virol. **64:**1091–1099.
21. **Persson, R., U. Svensson, and E. Everitt.** 1983. Virus receptor interaction in the adenovirus system. II. Capping and cooperative binding of virions on HeLa cells. J. Virol. **46:**956–963.
22. **Persson, R., C. Wohlfart, U. Svensson, and E. Everitt.** 1985. Virus receptor interaction in the adenovirus system: characterization of the positive cooperative binding of virions on HeLa cells. J. Virol. **54:**92–97.
23. **Philipson, L.** 1967. Attachment and eclipse of adenovirus. J. Virol. **1:**868–875.
24. **Philipson, L., E. Everitt, and K. Lonberg-Holm.** 1976. Molecular aspects of virus-receptor interaction in the adenovirus system, p. 203–216. In R. F. Beers, Jr., and E. G. Bassett (ed.), Cell membrane receptors for viruses, antigens and antibodies, polypeptide hormones, and small molecules. Raven Press, New York.
25. **Prage, L., U. Pettersson, S. Höglund, K. Lonberg-Holm, and L. Philipson.** 1970. Structural proteins of adenoviruses. IV. Sequential degradation of the adenovirus type 2 virion. Virology **42:**341–358.
26. **Salesse, R., and J. Garnier.** 1984. Adenylate cyclase and membrane fluidity. Mol. Cell. Biochem. **60:**17–31.

27. **Scatchard, G.** 1949. The attractions of proteins for small molecules and ions. Ann. N.Y. Acad. Sci. **51**:660–672.
28. **Schlegel, R., R. B. Dickson, M. C. Willingham, and I. H. Pastan.** 1982. Amantadine and dansylcadaverine inhibit vesicular stomatitis virus uptake and receptor-mediated endocytosis of α_2-macroglobulin. Proc. Natl. Acad. Sci. USA **79**:2291–2295.
29. **Sussenbach, J. S.** 1967. Early events in the infection process of adenovirus type 5 in HeLa cells. Virology **33**:567–574.
30. **Svensson, U.** 1985. Role of vesicles during adenovirus 2 internalization into HeLa cells. J. Virol. **55**:442–449.
31. **Svensson, U., and R. Persson.** 1984. Entry of adenovirus 2 into HeLa cells. J. Virol. **51**:687–694.
32. **Tomonaga, A., M. Hirota, and R. Snyderman.** 1983. Effect of membrane fluidizers on the number and affinity of chemotactic factor receptors on human polymorphonuclear leukocytes. Microbiol. Immunol. **27**:961–972.
33. **Wohlfart, C., and E. Everitt.** 1985. Assessment of specific virus infectivity and virus neutralization by a progeny virus immunotitration method as exemplified in an adenovirus system. J. Virol. Methods **11**:241–251.
34. **Wohlfart, C. E. G., U. K. Svensson, and E. Everitt.** 1985. Interaction between HeLa cells and adenovirus type 2 virions neutralized by different antisera. J. Virol. **56**:896–903.

Prospects for Antiviral Agents Which Modify the Pathway of Infection by Enveloped Viruses

ARI HELENIUS, MARGARET KIELIAN, JUDY WHITE,† and
JÜRGEN KARTENBECK

*Department of Cell Biology, Yale University School of Medicine, New Haven,
Connecticut 06510*

Although it is becoming increasingly clear that many viruses
depend on cellular functions in their entry, antiviral strategies must
focus on the virally determined steps. These include the attachment
to the cell surface and the uncoating reaction, which with enveloped
animal viruses usually involves a fusion reaction between the viral
envelope and a cellular membrane. The fusion factors in many
viruses are triggered by the mildly acid pH in endosomes. During
the past few years, a great deal has been learned about this fusion
reaction using in vitro assays and cultured cells. Some of the results
from the influenza and Semliki Forest virus systems suggest potential ways to inhibit entry. For example, fusion of Semliki Forest
virus requires cholesterol (or other steroids with a 3-β-OH group) in
the target membrane. This requirement is associated with an acid-
induced conformational change in the spike glycoprotein E1, which
may expose a cholesterol binding site at acid pH. This specific
interaction provides one possibility to inhibit entry with specific
drugs. Agents which modify the internal pH and other properties of
endosomes constitute another possibility. As the molecular details
of the penetration reaction for different virus species unfold, specific
targets for antiviral strategies are likely to emerge.

It is becoming increasingly clear that virus entry into host cells depends on
both cellular and viral functions. Not only do viruses rely on cellular surface
molecules for attachment, but many have to be internalized by endocytosis to
be infective. The details of the interactions between virus and cell during the
early stages of infection may suggest new strategies for future antiviral agents.

In this short paper we discuss some possible approaches. The ideas presented
are speculative and serve mainly to illustrate possible directions for this field of
research. The examples chosen are taken from the work in our laboratory on
togaviruses and myxoviruses in cultured cells. Both of these virus families enter
cells by membrane fusion in the endosome; i.e., they rely on endocytic uptake
and their fusion activity is triggered by the mildly acidic pH in the endosomal
compartment. These and many other viruses thus function as Trojan horses,
capitalizing on the uptake mechanisms of their hosts and penetrating at a
defined signal (low pH) only when already carried deep inside the cell. (For

†Present address: Department of Pharmacology, University of California-San Francisco, San
Francisco, CA 94143.

recent reviews on this mode of virus entry see references 1, 13, 15, 26; M. C. Kielian and A. Helenius, *in* H. Fraenkel-Conrat and R. Wagner, ed., *The Viruses*, in press; and White et al., this volume.)

To block entry of viruses, one may either modify the cellular processes on which it depends or block the activities of the viral components themselves. The challenge is twofold. Many of the cellular functions involved (such as receptor-mediated endocytosis through coated pits) may be crucial for the cell and thus unsuitable as targets for antiviral strategies. On the other hand, the steps in which the virus particles themselves play an active role are few in number and to date only partially understood in molecular terms. The sequence of main events culminating in infection may be summarized as adsorption, internalization, and penetration, and this is the order we follow in this discussion.

ADSORPTION

The initial step of infection involves virus binding to the cell surface. While this attachment does not automatically guarantee successful entry, it is a necessary step. Cells lacking the proper receptor are not infected or are infected very inefficiently. The specificity of binding varies greatly between viruses as described in other papers in this volume; some viruses display binding to a wide variety of cell types whereas others are highly restricted. In enveloped animal viruses this specificity depends on the viral spike proteins, and the cellular receptors in most cases appear to be glycoproteins, some of which have already been identified as functionally important to the host cell. In some cases (i.e., flaviviruses) the viruses are able to bind effectively only when complexed with antibodies because this allows them to utilize existing antibody receptors (Fc receptors) on the cell surface (21, 22). Binding frequently seems to rely on multiple virus-cell surface contacts in which each individual interaction is rather weak (6, 14). The avidity of binding is thus derived from the multivalency of the interaction.

To block the binding step, one may interfere with either the viral binding factors or the cellular receptors. Several types of interfering substances can be envisioned, and some have already been successfully used. Blocking agents directed against the surface receptors include: (i) antireceptor antibodies and their derivatives, which may be monoclonal or polyclonal, anti-idiotypic or idiotypic, and (ii) virus spike proteins or their analogs. The isolated viral spike glycoproteins have been shown to bind to the receptors and may, if presented in a suitable multivalent form (i.e., reconstituted vesicles or virus protein rosettes), prevent virus attachment competitively (6, 16).

Conversely, agents such as antibodies or receptor analogs directed against the spike proteins could be used. When the relevant receptor structures or determinants are known, we hope that it will be possible to devise a variety of agents which react specifically with the sites on the spike glycoprotein involved in receptor binding.

The surface molecules which serve as viral receptors obviously carry out functions in the cell independent of their inadvertent role in virus binding. Many may be receptors for physiological ligands and, as such, subject to

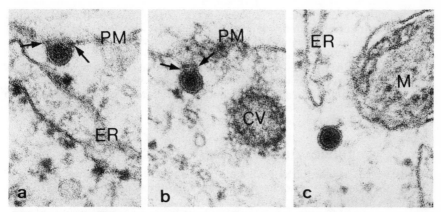

FIG. 1. Internalization of simian virus 40 particles by uncoated vesicles in CV-1 cells. The viruses appear to bind very tightly to the plasma membrane (PM) and then enter the cell in a tight-fitting plasma membrane vesicle devoid of a clathrin coat. The process morphologically resembles a budding process (see arrows) except that it occurs into the cell rather than out of the cell. ER, Endoplasmic reticulum; CV, coated vesicle; M., mitochondria. Magnification, ×133,300.

stringent control by the cell. In individual cases it may be possible to modify the virus-binding properties and expression of the receptor by the addition of the physiological ligands (agonists and antagonists). For example, receptors may be down-regulated by natural or artificial ligands. One well-studied case of down-regulation is the Fc receptor in macrophages (19, 20). The receptor carrying bound immune complexes is internalized by coated vesicles into the endosomal compartment and routed to the lysosome for degradation. In this case it has been demonstrated that receptor down-regulation depends on ligand valency and that receptors are sorted to lysosomes or back to the cell surface in endosomes (20). It is of interest that the Fc receptor is implicated in flavivirus infections. To our knowledge, no studies have yet been carried out on the effects of receptor down-regulation on infectivity of these viruses.

When manipulating the expression and functions of the cellular virus receptors, a major concern will be whether important cellular functions are affected as well. The severity of side effects must obviously be assessed in each individual case.

ADSORPTIVE ENDOCYTOSIS

The endocytic uptake of viruses occurs by adsorptive endocytosis. Electron microscopic analysis has indicated that some virus particles and virus families enter exclusively by clathrin-coated pits (2, 8). Others, such as influenza virus and simian virus 40, can be seen to enter, in addition, by close-fitting vesicles devoid of a coat (5; Fig. 1). Large viruses such as vaccinia virus may, in addition, enter by phagocytic, rather than pinocytic, processes. Whether all these forms of endocytosis lead to productive infection is not clear. Internalization of Semliki Forest virus involves coated vesicles and the standard intracellular pathway of receptor-mediated endocytosis (8, 15, 17). It relies on a constitutive, ongoing cellular process. While a number of conditions will inhibit receptor-mediated endocytosis in cultured cells, no treatment has yet been described that blocks uptake without the potential for serious cell damage upon prolonged treatment. It is, in our opinion, questionable whether cells and organisms are able to survive with inhibited endocytic functions. Therefore, the endocytic uptake step may not be a useful target for antiviral agents.

ACIDIFICATION OF THE ENDOCYTIC VACUOLES

Once internalized into the vacuolar apparatus of the cell, the penetration of viruses that depend on acidic pH in the endosomal compartment (25) can usually be inhibited by the addition of agents which elevate vacuolar pH. Such agents include lysosomotropic weak bases such as ammonium chloride, chloroquine, tributylamine, methylamine, and other lipophilic amines (9, 10). Amantadine and its homologs are also lysosomotropic weak bases, and our studies with Semliki Forest virus clearly indicate that this property can block entry (12). Whether the same mechanism applies to its prophylactic effect on influenza A infection remains unclear at this point.

Another group of agents which elevate the vacuolar pH, albeit by a different mechanism, are carboxylic ionophores such as monensin and nigericin. Our studies have shown that these agents resemble lysosomotropic weak bases in having only a marginal effect on the uptake of bound viruses by endocytosis, on the transport of viruses to endosomes and lysosomes, and on most of the biosynthetic steps following penetration (18). However, they eliminate the acid conversion of the virus into the form active in fusion and inhibit penetration (8a; M. C. Kielian and A. Helenius, manuscript in preparation).

Although effective in blocking virus entry in cultured cells, most of the agents that affect endosomal pH may, with the possible exception of amantadine, be too harmful at the concentrations needed to find extensive use as antiviral agents. Lysosomotropism (i.e., accumulation into lysosomes and other acid organelles) may, however, turn out to be a useful method in the cellular targeting of antiviral agents which themselves act by some other principle (3).

The H^+-ATPase responsible for the acidification of endosomes constitutes another cellular target against which antiviral strategies may be directed. It is now clear that the enzyme belongs to a group of ATPases which are found in several endocytic and exocytic organelles (G. Rudnick, in T. E. Andreoli, D. D. Fanestil, J. F. Hoffman, and S. G. Schultz, ed., Physiology of Membrane Disorders, 2nd ed., in press). It is electrogenic, inhibited by N-ethyl maleimide, and insensitive to vanadate, oligomycin, and ouabain (7). However, since the enzyme has not yet been isolated and characterized, its detailed properties are not known. As more is learned about this important enzyme, it may become possible to devise specific inhibitors which would result in endosomal pH elevation. Ideally, such inhibitors should be coupled to ligands specific for the virus receptor, thus targeting the agent only to virus-susceptible cells.

VIRAL FUSION ACTIVITY

Penetration by enveloped viruses requires a membrane fusion event induced by viral spike glycoproteins. Depending on the pH threshold, the reaction may occur either on the plasma membrane or in the acidic vacuoles of the endocytic pathway. The data from several virus systems suggest that fusion is concomitant with specific conformational changes in the viral fusion proteins and that these changes are induced by acid pH (4, 11a, 24; White et al., this volume). It is this conformational switch which is inhibited when endosomal pH is elevated. Although much is already known about the nature of the conformational change in the influenza virus hemagglutinin, useful strategies for blocking the reaction directly have not emerged. Hydrophobic peptides with amino acid sequences similar to the postulated fusion peptide in influenza virus hemagglutinin have been tested, but their effects on fusion per se remain inconclusive (23).

In the case of togaviruses (alphaviruses) such as Semliki Forest virus and Sindbis virus, the acid-activated fusion requires the presence of a sterol such as cholesterol in the target membrane (reviewed in reference 26). The specificity is strictly limited to sterols with a 3-β-hydroxyl group (11). The specificity of the reaction as determined by analysis of various cholesterol analogs does not correlate directly with any known effect of cholesterol on bilayer structure. Recently, studies in our laboratory have shown that this requirement is a property of the E1 glycopolypeptide chain of the heterotrimeric spike protein, which can undergo its acid-induced conformational change only if cholesterol is present (11a). Since fusion in this case depends on the presence of compounds with a highly stereo-specific structure, it would seem feasible to devise specific inhibitors against this group of viruses. The cholesterol dependence could constitute a starting point for a highly focused search for inhibitors of togaviruses.

CONCLUSIONS

As the details of the entry pathway followed by different enveloped and nonenveloped viruses are unraveled further, targets for antiviral approaches may become apparent. Viruses use varied receptors, and their fusion proteins are different in structure and function; it will therefore be important that studies be pursued in parallel on all major virus families. A common drug or strategy that will block the entry of all enveloped viruses seems unlikely. The focus should be on the molecular events involved, while the role of the biological aspects of the cell should not be overlooked.

The work was supported by Public Health Service grants AI18582 and AI18599 from the National Institutes of Health and by the Swebelius Foundation.

LITERATURE CITED

1. **Brown, M. S., R. G. W. Anderson, and J. C. Goldstein.** 1983. Recycling receptors: the round-trip itinerary of migrant membrane proteins. Cell **32**:663–667.
2. **Dales, S.** 1973. Early events in cell-animal virus interactions. Bacteriol. Rev. **37**:103–135.
3. **De Duve, C., T. De Barsy, B. Poole, A. Trouet, P. Tulkens, and F. Van Hoof.** 1974. Lysosomotropic agents. Biochem. Pharmacol. **23**:2495–2531.

4. **Doms, R. W., A. Helenius, and J. White.** 1985. Membrane fusion activity of the influenza virus hemagglutinin: the low pH-induced conformational change. J. Biol. Chem. **260:**2973–2981.

5. **Dourmashkin, R. R., and D. A. J. Tyrrell.** 1974. Electron microscopic observations on the entry of influenza viruses into susceptible cells. J. Gen. Virol. **24:**129–141.

6. **Fries, E., and A. Helenius.** 1979. Binding of Semliki Forest virus and its isolated glycoproteins to cells. Eur. J. Biochem. **97:**213–220.

7. **Galloway, C. J., G. E. Dean, M. Marsh, G. Rudnick, and I. Mellman.** 1983. Acidification of macrophage and fibroblast endocytic vesicles *in vitro.* Proc. Natl. Acad. Sci. USA **80:**3334–3338.

8. **Helenius, A., J. Kartenbeck, K. Simons, and E. Fries.** 1980. On the entry of Semliki Forest virus into BHK-21 cells. J. Cell Biol. **84:**404–420.

8a.**Helenius, A., M. Kielian, J. Wellsteed, I. Mellman, and G. Rudnick.** 1984. Effects of monovalent cations on Semliki Forest virus entry into BHK-21 cells. J. Biol. Chem. **260:**5691–5697.

9. **Helenius, A., M. Marsh, and J. White.** 1982. Inhibition of Semliki Forest virus penetration by lysosomotropic weak bases. J. Gen. Virol. **58:**47–61.

10. **Jensen, E. M., and O. C. Liu.** 1961. Studies of inhibitory effect of ammonium ions in several virus-tissue culture systems. Proc. Soc. Exp. Biol. Med. **107:**834–838.

11. **Kielian, M. C., and A. Helenius.** 1984. Role of cholesterol in fusion of Semliki Forest virus with membranes. J. Virol. **52:**281–283.

11a.**Kielian, M., and A. Helenius.** 1985. pH-induced changes in viral glyco-proteins involved in the fusion activity of Semliki Forest virus. J. Cell. Biol. **101:**2284–2291.

12. **Kielian, M., S. Keranen, L. Kaariainen, and A. Helenius.** 1984. Membrane fusion mutants of Semliki Forest virus. J. Cell Biol. **98:**139–145.

13. **Lenard, J., and D. Miller.** 1982. Uncoating of enveloped viruses. Cell **28:**5–6.

14. **Lonberg-Holm, K., and L. Philipson.** 1974. Early interactions between animal viruses and cells. Monogr. Virol. **9:**1–148.

15. **Marsh, M.** 1984. The entry of enveloped viruses into cells by endocytosis. Biochem. J. **218:**1–10.

16. **Marsh, M., E. Bolzan, J. White, and A. Helenius.** 1983. Interaction of Semliki Forest virus spike glycoprotein rosettes and vesicles with cultured cells. J. Cell Biol. **96:**455–461.

17. **Marsh, M., and A. Helenius.** 1980. Adsorptive endocytosis of Semliki Forest virus. J. Mol. Biol. **142:**439–454.

18. **Marsh, M., J. Wellsteed, H. Kern, E. Harms, and A. Helenius.** 1982. Monensin inhibits Semliki Forest virus penetration into baby hamster kidney (BHK-21) cells. Proc. Natl. Acad. Sci. USA **79:**5297–5301.

19. **Mellman, I., and H. Plutner.** 1984. Internalization and fate of macrophage Fc receptors bound to polyvalent immune complexes. J. Cell Biol. **98:**1170–1177.

20. **Mellman, I., H. Plutner, and P. Ukkonen.** 1984. Internalization and rapid recycling of mouse macrophage Fc receptors bound to monovalent antireceptor antibody, possible role of a prelysosomal compartment. J. Cell Biol. **98:**1163–1169.

21. **Peiris, J. S. M., S. Gordon, J. C. Unkeless, and J. S. Porterfield.** 1981. Monoclonal anti-Fc receptor IgG blocks antibody enhancement of viral replication in macrophages. Nature (London) **289:**189–191.

22. **Peiris, J. S. M., and J. S. Porterfield.** 1979. Antibody mediated enhancement of flaviviruses replication in macrophage-like cell lines. Nature (London) **282:**509–511.

23. **Richardson, C., A. Scheid, and P. Choppin.** 1980. Specific inhibition of paramyxo-virus and myxovirus replication with oligopeptides with amino acid sequences similar to those at the N-termini of the F, or HA₂ viral polypeptides. Virology **105:**205–222.

24. **Skehel, J., P. Bayley, E. Brown, S. Martin, M. Waterfield, J. White, I. Wilson, and D. Wiley.** 1982. Changes in the conformation of influenza virus hemagglutinin at the pH optimum at virus-mediated membrane fusion. Proc. Natl. Acad. Sci. USA **79:**968–972.

25. **Tycko, B., and F. B. Maxfield.** 1982. Rapid acidification of endocytic vesicles containing α2-macroglobulin. Cell **28:**643–651.

26. **White, J., M. Kielian, and A. Helenius.** 1983. Membrane fusion proteins of enveloped animal viruses. Q. Rev. Biophys. **16:**151–195.

Author Index

Allaway, Graham P., 116
Almond, J. W., 28

Bachi, Thomas, 60
Bassel-Duby, Rhonda, 13
Baxt, Barry, 126
Burness, Alfred T. H., 116

Callahan, Pia L., 109
Caton, Andrew, 60
Co, Man Sung, 138
Cohen, Gary H., 74
Colonno, Richard J., 21, 109
Consigli, R. A., 44
Cooper, Neil R., 160
Crowell, Richard L., 1, 103

Doms, Robert, 54

Eisenberg, Roselyn J., 74
Evans, D., 28
Everitt, Einar, 196

Ferguson, M., 28
Fields, Bernard N., 13, 138
Fitzgerald, David, 191

Gaulton, Glen N., 138
Gerhard, Walter, 60
Gething, Mary-Jane, 54
Greene, Mark I., 138
Griffith, G. R., 44

Helenius, Ari, 54, 205

Jayasuriya, Anula, 13

Kartenbeck, Jürgen, 205
Kielian, Margaret, 54, 205
Krah, David L., 103

Landsberger, Frank R., 85
Lenard, John, 182
Liu, Jing, 138
Lonberg-Holm, Karl, 1
Long, William J., 109
Ludlow, J. W., 44

McClintock, Patrick R., 36
Madshus, Inger Helene, 171
Mapoles, John E., 103
Marriott, S. J., 44
Martin, Elaine, 144
Minor, P. D., 28
Morgan, Donald O., 126
Mosser, Anne, 21
Murayama, Jun-Ichiro, 144

Nemerow, Glen R., 160
Notkins, Abner Louis, 36

Oliff, Allen, 91
Olsnes, Sjur, 171

Pardoe, Ingrid U., 116
Pastan, Ira, 191
Paulson, James C., 144
Persson, Robert, 196
Prabhakar, Bellur S., 36

Reagan, Kevin J., 152
Rogers, Gary N., 144
Rossmann, Michael, 21
Rueckert, Roland, 21

Sandvig, Kirsten, 171
Schild, G. C., 28
Schlegel, Richard, 66
Sehgal, Pravinkumar B., 85
Seth, Prem, 191
Sherry, Barbara, 21
Siaw, Martin F. E., 160
Svensson, Ulla, 196
Sze, Gloria, 144

Tavakkol, Amir, 116
Taylor, Alex, 60
Tomassini, Joanne E., 109

White, Judy, 54, 205
Willingham, Mark, 191
Wohlfart, Claes, 196
Wunner, William H., 152

Yewdell, Jonathan W., 60

Subject Index

Acetylcholine receptor, 152
Acridine orange-sensitized virus, 171
Adenovirus
  attachment, 196
  effects of low pH on, 191
  penetration, 191, 196
  uncoating in vitro, 196
  virion attachment protein, 191, 196
Amantadine-resistant mutants, 54
Antiviral agents, 205
Anti-idiotype antibodies, 1, 28, 36, 126, 138

Beta-adrenergic receptor, 138
Bromelain-released hemagglutinin, 60

Cellular receptors for C3, 160
Cellular receptors for viruses
  adenoviruses, 191, 196
  anti-idiotype antibodies
    lack of recognition, 1, 28, 36, 126
    recognition, 1, 44, 138
  cardioviruses, 116
  complexed with virus, 103
  coxsackievirus type A, 109
  coxsackievirus type B, 1, 36, 103
  Epstein-Barr virus, 160
  encephalomyocarditis virus, 116
  flaviviruses, 205
  foot-and-mouth disease virus, 126
  general, 1, 138, 205
  influenza virus, 116, 144
  polyomavirus, 44
  poliovirus, 28
  rabies virus, 152
  reovirus, 13, 138
  rhinovirus, 21, 109
  Sendai virus, 144
  specificity, selective pressure, 144
  vesicular stomatitis virus, 182
  virus mutants, lack of, 28, 109
Cellular receptors, isolation
  coxsackievirus type B, 1, 103
  Epstein-Barr virus, 160
  general, 1
  polyomavirus, 44

212

rhinovirus, 109
    reovirus, 138
Colloidal gold in study of virus penetration, 191
Cooperative effects, adenovirus attachment and uncoating, 196
Clathrin-coated pits, 1, 182, 191, 205
Cross-linking agent, photoreactive, 44
C3, cellular receptors, 160

DEAE-dextran, enhanced virus attachment, 152, 182
Differentiation of cellular receptors for rabies virus, 152
Discontinuous epitopes, glycoprotein D, 74

Endosomes, site of penetration, 1, 54, 171, 182, 191, 205
Encephalomyocarditis virus
    attachment to erythrocytes, 116
    hemagglutination, 116
    pH required for penetration, 171
Enhancers, leukemogenic potential, 91
env gene, leukemia virus, 91
Epstein-Barr virus
    diseases, 160
    mitogenic activity, 160
    transformation of B cells, 160
Erythrocyte
    modification of surface, 116, 144
    receptors
        C3 (CR2), 160
        encephalomyocarditis virus, 116
        influenza virus, 116, 144
Erythroleukemia, murine, 91

"Fluidizers," membrane, 196
Foot-and-mouth disease virus, effect of trypsin, 126
Friend murine leukemia virus complex, 91
Fusion proteins, virus, 54, 60, 66, 85, 182, 205
    sequence homology, 85
Fusion peptides, 54, 205

G protein, vesicular stomatitis virus, 66, 182
Glycophorin
    isolation, 116
    receptor for viruses, 116, 144
    structural modification, 116, 144
Glycoprotein D, discontinuous epitopes, 74

Hemolysis
    by peptides, 66
    by vesicular stomatitis virus, 66, 182

Herpesvirus glycoprotein D, 74
Host-mediated selection of receptor variants, 144

Influenza virus
  attachment to erythrocytes, 116
  cDNA, 54
  replication in avian and human cells, 144
  hemagglutinin, 54, 60, 85, 116, 144
  maturation, 60
  mutants, 144
  penetration by fusion, 54
  pH variants, 54, 85
  receptor-specific variants, 144
Inhibitors of endosomal acidification, 171, 182, 205
"Internal image" of virion attachment protein, 36, 138
Ion permeability, effect on G protein, 66
Ionophore, inhibition of virus penetration, 171, 205

Leukemia virus
  envelope gene, 91
  recombinant, 91
Lipid transport protein, 85
Liposomes, 54, 66, 144, 191
Lysosomotropic amines, 183, 205

Membrane fusion, 54, 85
Mink cell focus-inducing virus, 91
MN blood group activity, 116
Monoclonal antibodies
  adenovirus penton, 191
  B-lymphocyte receptors for Epstein-Barr virus, 160
  coxsackievirus type B, 36
    cellular receptors, 1
  foot-and-mouth disease virus, 126
  herpesvirus glycoprotein D, 74
  influenza virus hemagglutinin, 60
  polyomavirus, 44
  poliovirus, 28
    cellular receptors, 28
  polypeptides, 28
  reovirus, 13, 138
  rhinovirus, 21, 109
    cellular receptors, 109
Monoclonal antibody-resistant mutants
  influenza virus, 60
  poliovirus, 28
  reovirus, 13
  rhinovirus, 21, 109

Myotubes, 152

Neutralization immunogen (NIm) groups, 21

Pathogenesis and virus tropism
  general, 1, 13
  leukemia virus, 91
  rabies virus, 152
  reovirus, 13
Phosphatidylcholine-specific transport protein, 85
Picornavirus penetration into cell, 171
Poliovirus
  A particle, 171
  antibody binding sites, 28
  penetration into cell, 171
  photosensitive, 171
Polyomavirus
  capsid proteins, 44
  posttranslational modification, 44
  hemagglutinin, 44
  empty capsids, 44
Polypeptides
  antibodies to, 28, 66, 74
  biological activity, 54, 66

Rabies virus
  cellular receptors, 152
  maturation, 152
  pathogenesis, 152
Receptor (see Cellular receptor)
Receptosome (see Endosome)
Reovirus
  cellular immunity, 13
  genetic reassortment, 13
  hemagglutinin, 13, 138
Retinol binding protein, receptor for, 138
Rhinovirus
  antibody binding sites, 21
  capsid structure, 21
  penetration into cell, 171
  photosensitive, 171

Semliki Forest virus
  fusion with cell membrane, 205
  lysosomotropic weak bases as inhibitors, 205
Sialic acids
  chemical modification, 116, 144
  in receptor determinants

for encephalomyocarditis virus, 116
for influenza virus, 116, 144
for rabies virus, 152
on receptor for reovirus, 138
removal enhances vesicular stomatitis virus attachment, 182
Sialyl transferase, treatment of receptors, 144

Toxins, entry into cell, 171, 191
Triton X-114, use with viruses, 171, 191

Vesicular stomatitis virus
glycoprotein (G protein), 6, 182
hemolysis by G protein, 66
lysosomotropic amines as inhibitors, 182
nonspecific attachment to cells, 182
Virion attachment protein
adenovirus fiber, 191, 196
general, 1
foot-and-mouth disease virus, 126
influenza virus hemagglutinin, 54, 60, 144
polyomavirus, 44
reovirus, 13, 138
rhinovirus, 21
vesicular stomatitis virus, 66, 182
Viropexis, 1
Virus
leukemia induction, 91
penetration, 1, 44, 54, 160, 171, 182, 191, 196, 205
receptor (*see* Cellular receptor)
uncoating, 1, 54, 182, 191, 196, 205
Virus-receptor complex, isolation, 103

X-ray crystallographic structure
influenza virus hemagglutinin, 54, 85
rhinovirus, 21